Practical Science for Gardeners

Mary M. Pratt

Practical Science for Gardeners

Mary M. Pratt

Timber Press

Contents

Preface

This book aims to be a little different from other gardening books. It might be best described as the 'appliance of science for gardeners' – but *simple* science, which I hope you'll find easy to understand. You won't discover in it the month-by-month calendar of essential tasks found in many other books and magazines; it isn't a guide to garden design or garden history; nor is it a technical horticultural textbook. It explains, in a straightforward way, what science can tell us about how plants are constructed and how they grow and flourish. It then applies this knowledge to everyday gardening tasks, to some of the common problems gardeners face, and to accepted practices, whether they be so-called 'organic' or otherwise. Equipped with a better knowledge of the structure of delicate plant tissues, and the way plants function and interact with their environment, we might have second thoughts about some of our gardening techniques. Perhaps I'll explode some myths. Above all I hope that I may challenge and entertain. As I will explain later, conclusive understanding of life processes is elusive, and there is a great deal we still don't know for certain. My gardening suggestions are therefore not to be swallowed whole as dogma – they might, instead, be pointers to your own observations and experiments.

Gardening is 'chemistry' – but don't panic! It's just that I believe such a notion aptly encompasses my philosophy of gardening as a perfect collaboration between emotion and rationality, between the feeling and reasoning parts of our psyches. We talk about the 'chemistry' of human relationships – of feeling 'on net' with someone, of 'falling in love' or of experiencing irritation or antagonism. In this sense, the proverbial green-fingered gardener is someone whose relationship with plants achieves the right 'chemistry' in an instinctive, 'seat-of-the-pants' sort of way. But there is another sense in which it is chemistry – for the growth of plants is all to do with the way in which chemical substances maintain and control life. If we start with an understanding of this, we're well on the way to meeting the needs of plants and creating a flourishing, beautiful and productive garden.

Artistic creativity and that instinctive feeling for the plant world are gifts of personality which I believe defy analysis. Maybe psychology will one day explain them but, since I am not a psychologist, I'm concerned largely with the other kind of chemistry and its close cousins, biology (the study of living organisms) and ecology (the study of living things in relation to their environment). If you're already naturally gifted, I hope this will add to your skills: if you despair of ever possessing 'green fingers', it may even help to generate more creativity. Rest assured that it will be *simple* science. What's more, I make no claim that science has all the answers – indeed, in the first chapter I suggest that science is largely misunderstood, and that rather than providing incontrovertible proof of anything it is, at best, only giving us *provisional* knowledge. I hope you'll agree, though, that it is knowledge worth having.

I am very grateful for help of various kinds from the following: Dr. Ken Thompson, Katharine Swift, Dr. Tony Polwart, Valerie Brooke (RHS – Royal Horticultural Society – Wisley Library), Norman Day (Forest Research), Blackwell Publishing, Hodder & Stoughton, Cambridge University Press, Oxford University Press, Faber and Faber, The Wildfowl Trust; Gwen Leighton and John Fulcher for illustrations; and Dr Keith Goodway and Professor Michael Page who kindly read sections in draft. I do, of course, nonetheless take full responsibility for all errors and omissions.

Last but not least I must thank my husband – for moral support, for technical assistance, and for many hours of help with the final editing.

Note: After Mary Pratt was diagnosed with a brain tumour, she worked with great determination, in various hospitals and then at home, to complete the text and illustrations for this book. She died in December 2004, delighted to know that the book had been accepted without question for publication.

She had briefed her husband, Professor George Pratt, about the non-scientific practical details which remained to be dealt with.

1. Muck, magic and molecules

This chapter is about the chemistry of life and about how scientists unravel the complex and variable processes that govern the lives of plants. But don't be alarmed – it's going to be a gentle introduction.

All living organisms are made of chemical elements. Plants depend on them for their survival and growth, whether they come from compost or a packet of 'Grow-Well'. A recently published book was described on the cover as a 'guide to natural and chemical-free gardening'. With due respect to much of the excellent content, gardening is, by definition, not natural, and to describe it as 'chemical-free' is misleading. We may choose to avoid using man-made chemical additives, but plants actually depend on chemistry to sustain the whole amazing world of their inner workings.

What science can tell us about the chemistry of life

By painstaking observation, then putting forward hypotheses and experimenting to test these hunches, scientists have discovered that our planet is constructed of around 100 chemical elements, each occurring in the form of tiny units – atoms – indivisible by ordinary physical or chemical means. (I say 'around' because a few of them have been created in the laboratory and do not occur naturally.) Atoms group themselves together into larger units called molecules, and it is molecules containing the element **carbon** which make up the structure of living organisms. Such molecules merit the term **'organic'**, but the only difference between them and the simpler molecules of a child's first chemistry set is that they are constructed around carbon atoms and tend to be rather large (Figure 1-1).

The architecture of these molecules, and the way in which they interact and fix themselves together to form a living organism, is truly stupendous. Likewise, the way in which waste products and dead organisms – in gardening parlance 'muck' – become broken down again into their constituents is equally miraculous. But it is nonetheless chemistry. The living world is a massive and incredibly complex DIY chemical-processing operation, thought to have started up somewhere between 3.5 and 4 billion years ago, and getting ever more sophisticated since.

Figure 1-1. The difference between a protein molecule and a molecule of common salt. The protein molecule is made up of a chain of sub-units (a a) each being an amino acid, of which there are 20 different kinds. Only a *very* small part of a protein molecule is shown here. A complete protein would be many times longer, and the chain is often twisted or folded into intricate shapes.

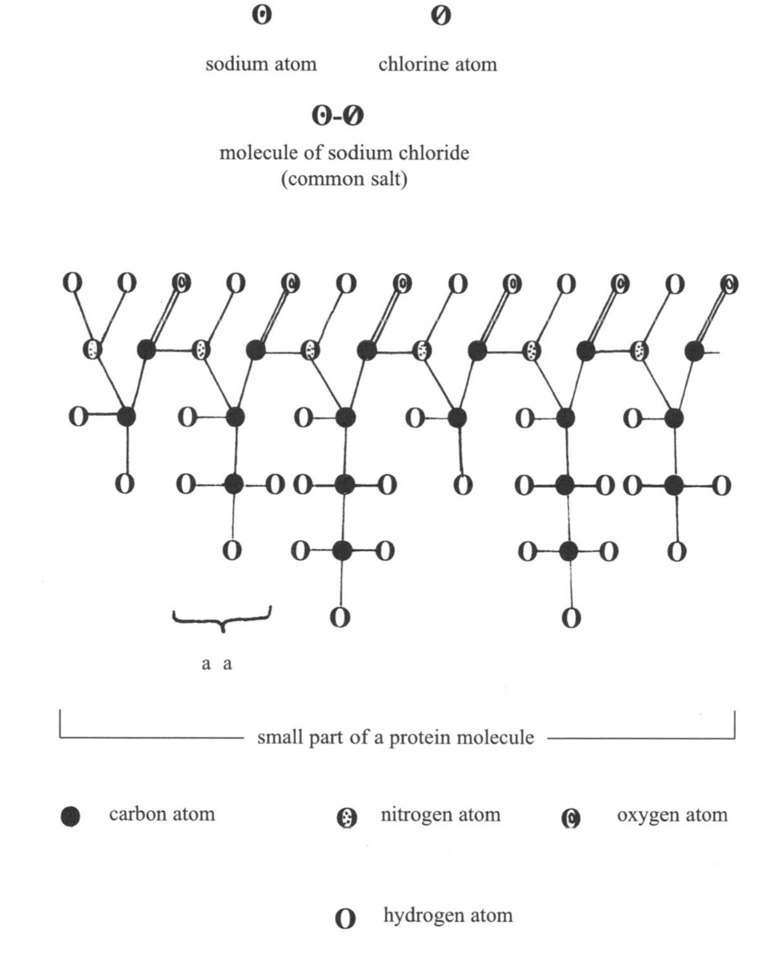

Living things are continually taking molecules apart and putting them together in different ways. When we're gardening we are truly playing with muck and magic, but in the form of a gigantic and amazing chemistry set.

Studying plants – the scientific approach and its limitations

Scientists have unravelled the way in which living things operate, and it is therefore to science that we should look for help in understanding how plants grow. Unfortunately living organisms are so complex and so variable (some would say 'bloody minded') that studying them is no simple matter. Moreover, science doesn't claim to give perfect and immutable *proof* for all time – it simply gives us provisional knowledge (which is better than guesswork or opinion). As in medical science things change as new information becomes available. Very often there are no straightforward answers (though gardening 'experts' would sometimes have us believe that there are).

Because of this complexity and variability, and the many interacting factors in the environment that affect plants, no individual plant will behave in exactly the same way as its neighbours. Scientific experiments therefore have to depend on statistics and can only come up with answers which are **probabilities**.

Consider an experiment to find out if a particular fertilizer is more effective than another. Two plots are chosen, and an effort made to render all environmental factors (soil type, exposure to light, drainage and so on) the same. Then 100 seedlings are planted in each plot and one treated with fertilizer A, the second with fertilizer B.

After an allotted time all the plants are measured for height and dry weight. (The weight of dried-out plant material gives a more realistic measure of actual plant tissue because cells contain large amounts of water, which may vary between individual plants.) To a casual observer the plants in plot A might appear to have done better, but in neither plot will the plants be uniform. There will be a range of measurements, due partly to the variability of living processes and partly to the fact that it is very difficult to make other environmental factors identical throughout each plot. You might suggest taking the average, and the fertilizer which gives the highest average height/weight can then be said to be the best. But what if in plot A – although the average is higher – the range of variation is

greater, so that there are a few very small plants, several enormous ones and lots of fairly unimpressive middling ones in between? You might be less inclined to conclude that fertilizer A was the best. To overcome this problem the results will be analysed using statistical tests to produce an answer which can only say that there is a certain **percentage probability** that fertilizer A is better than B. Statisticians talk about there being a **significant difference** between them at a certain level of probability.

If the experiment is repeated, the chances are that the results will be slightly different even though an attempt has been made to keep the conditions the same. I recollect an experiment, carried out on a popular TV programme, in which different kinds of hanging basket materials were tested. Though the conditions were ostensibly the same, only *one* of each type of basket was planted up. A winner was chosen by visual examination, but I wonder what the results would have been had *several* of each been tried out and the procedure repeated – a much fairer test. Even then, I suspect that the results would not have been clear-cut.

I use these illustrations to emphasize that biological science – and indeed science in general – gives us the best possible answers or explanations, in the circumstances, and for the time being. But they are only provisional. All it can do is say that the evidence supports a particular hypothesis and that conclusions are consistent with other findings – in other words, the story seems to *make sense.*

Having said that, there are some aspects of the life of plants about which we can be pretty certain – particularly when considering directly observable internal structures, and intensively studied universal functions, such as the way plants make food by photosynthesis and release energy by respiration. Here we come back to molecules because, whether we are made of ‘sugar and spice and all things nice’ (girls) or ‘frogs and snails and puppy dogs’ tails’ (boys), we are all, in the last resort, constructed from proteins, fats and carbohydrates, unromantic as that may seem!

The rest of this chapter takes a look inside a plant to see what it is made of and what goes on there. There is, of course, no such thing as a ‘typical’ plant; also, because living processes are extremely complex, I have had to simplify and generalize. But there are things which are relevant to gardeners which are common to all green plants. Understanding them can only help our gardening.

The building blocks

Most plants and animals are made up of millions of microscopic building blocks or **cells**. In plants each cell has a non-living cell wall, typically made of cellulose. Inside this is a ‘dollop’ of what appears under an ordinary light microscope to be a jelly-like material, the **cytoplasm**.

Figure 1-2. Generalized diagram of two adjacent plant cells, showing the main features. (Most plant cells are in the order of hundredths of a millimetre in diameter, so this is hugely magnified.) The cell wall is made of tough cellulose and, to make communication between cells easier, there are usually tiny pores through which a strand of cytoplasm passes. Cells come in many different shapes and sizes according to where they occur in the plant and the different jobs they have to do.

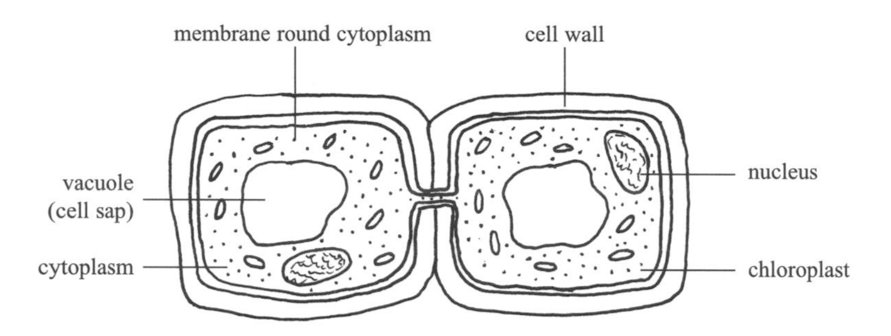

The cytoplasm is covered by a thin 'skin' which allows water and certain other small molecules through, and a similar membrane lines any holes (or vacuoles) within the cytoplasm. At certain places in the cell wall there may be strands of cytoplasm which pass through and connect with a neighbouring cell. On much closer examination – with a more sophisticated electron microscope – the cytoplasm turns out to contain sheets and sheets of membranes lying in a watery solution of chemicals. We now know that these structures, including the outer cell membrane, are made up of a double layer of **fat (lipid)** molecules sandwiched between two layers of **protein** molecules. The same applies to animal cells – so humans and plants are both made largely of protein and fat. Plants are a bit different, in having a 'skeleton' of cell walls made of the carbohydrate, cellulose.

The structure labelled the **nucleus** is the control centre of the cell and it is here that the genetic material, the chemical compound **DNA (deoxyribonucleic acid)**, is situated. Its role is to send messages, encoded in a similar chemical substance **RNA (ribonucleic acid)**, out into the cytoplasm giving instructions for specific chemical reactions to be carried out (see Chapter 7). These chemical reactions determine what sort of plant it is and how it germinates, grows and reproduces.

Staying alive and growing – food and energy

The main feature distinguishing plants from animals is the ability of the former to stay rooted to the spot, while animals have to scurry around, searching for nourishment. Plants make their own food using simple chemicals from air and soil, whereas animals have to rely on food already made for them, not necessarily near at hand. But before this food can be put to use for growing new parts, energy has to be made available – no factory can operate without a source of power.

In the human economy a major source has been the burning of fossil fuels; in the plant economy power comes from burning sugar – the process of **respiration** inside cells, actually a very slow kind of combustion (but without fire and smoke).

The source of power – respiration

Respiration has important implications, particularly for greenhouse gardening. Like houses, all plants are constructed of building blocks or cells. But, *unlike* houses, each cell is a miniature factory. It takes in raw materials and churns out products, which may be destined for growth by making more cells, or may be waste, which needs to be disposed of. Respiration is the energy-supply process. It is going on in all cells all the time, making energy available for carrying out the numerous tasks which keep an organism alive and help it to grow.

Energy is stored in sugar, which in turn can be stored in the form of starch – often in special storage organs underground (hence starchy potatoes). For the energy to be released the sugar is very, very slowly 'burnt' using oxygen; glucose is made to react with oxygen inside special 'power stations' called mitochondria. The sugar is broken down, energy is temporarily packed away in an easily accessible form, whilst the by-products – carbon dioxide and water – are disposed of or used elsewhere. Detailed experiments have shown that the whole business of respiration is actually a complicated chain of chemical reactions facilitated by specialist proteins called **enzymes**. (Enzymes are like the catalysts of test-tube chemistry – they help reactions along; see Chapter 3 where their role in germination is explained.)

Another by-product is, of course, heat – very noticeable in our own bodies when we are frantically active, and in compost heaps where bacteria are busy (see Chapter 4). Heat production is not so noticeable in plants because, being immobile, they use a lot less energy. Having said that, some plants *are* able to raise their temperature by a surprising amount; one example which has been studied recently is lords and ladies (*Arum maculatum*). This extra heat comes from respiration.

I have described respiration as a process of slow combustion in the presence of oxygen, but there are exceptions. Animal muscle cells and some micro-organisms (bacteria and yeasts) are capable of releasing energy from carbohydrate without the presence of oxygen and, instead of water, the end product – in addition to carbon dioxide – may be either lactic acid or alcohol. This process of **anaerobic respiration** is not as efficient, but we have benefited from it by being able to use yeast to ferment various plant extracts (and enjoy the end result!).

Figure 1-3. Simplified diagram of the internal structure of a cell (not to scale), showing the functions of the different parts. There are generally lots of mitochondria and chloroplasts in a single cell, but usually only one Golgi body and one nucleus. The membranes with tiny projections on them – the ribosomes – extend throughout the cell. All these tiny internal structures (or organelles) have been revealed by the use of the electron microscope, which gives images magnified hundreds of thousands times (for example x 250,000) in contrast to the ordinary light microscope which can only give a maximum magnification of x 1,500. A typical mitochondrion is in the order of one thousandth of a millimetre in diameter.

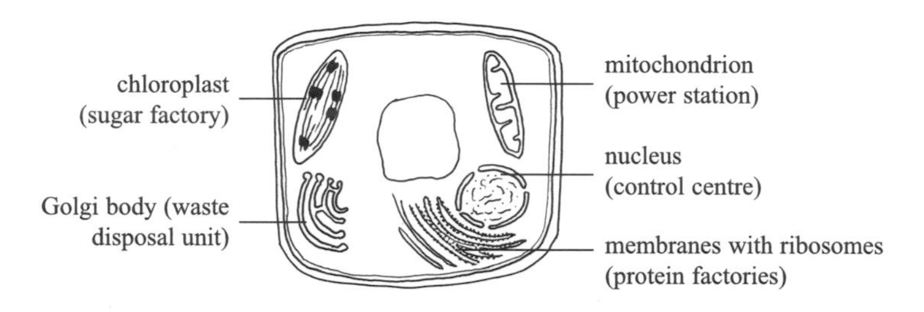

One important use for the energy which respiration makes available is to empower the construction of the giant proteins and other large molecules which are the structural components needed for growth – cellulose for cell walls, lignin (another kind of strong carbohydrate) for strengthening stems. They are manufactured on 'assembly lines' under instructions from the control centre, the nucleus, and are transported through the cell along channels and out to cells in other parts of the plant, particularly the growing points.

Exactly the same thing is happening all the time in our own cells and, of course, we breathe out the carbon dioxide. For us, the inhalation of oxygen/getting rid of carbon dioxide is called breathing, but scientific studies have shown that the actual energy-releasing process of respiration is much the same in plants and animals.

For gardeners, a particularly relevant point is that respiration is very sensitive to temperature – the higher the temperature the more rapid the reaction. In a heated greenhouse it may speed up so much that precious glucose is being wasted (see the discussion of the balance between photosynthesis and respiration, page 15).

Making materials – photosynthesis

So far so good – but where does the sugar come from, together with all the other chemicals needed to manufacture proteins and other plant materials? Plants can't eat other living things to obtain them (insectivorous plants are notable exceptions) so they have either to *make* them, or obtain them from the soil.

They solve the problem by making sugar by the process of **photosynthesis**. They subsequently have to make other things, but the food with which every plant starts is sugar (glucose). This is made *in situ* in the leaves by a remarkable, solar-powered and extremely complicated series of reactions – so remarkable that (although they reckon they are getting ever nearer to a complete understanding) biochemists have not yet replicated the process from scratch. Anyone who could achieve photosynthesis 'in a test tube' would save the planet and leave multi-billionaire Bill Gates standing.

Photosynthesis is the exclusive ability of green plants to manufacture sugar from carbon dioxide and water and, in so doing, to trap the sun's energy, ultimately making it available to other living creatures. In the cells of the green parts of a plant – most importantly in the leaves – there are tiny photosynthetic units, the **chloroplasts**, where the light is absorbed with the help of the green pigment, **chlorophyll**, and stored up as chemical energy in glucose.

The relationship between plants and light is thus of the utmost importance to gardeners. Some plants need more than others; leaves are specially designed as solar power houses but need careful treatment if they are to function optimally (see Chapter 2). You'll notice that I've been talking about *green* plants. Plants with reddish or brownish leaves have slightly different photosynthetic pigments, which do the same job as chlorophyll.

As well as being light-dependent, the two other determinants of photosynthetic efficiency are temperature and carbon dioxide concentration. It is no good having plenty of light if there is insufficient carbon dioxide or the temperature is too low. These are important factors in greenhouse management. Remember also that we have just noted that respiration (the *opposite* of photosynthesis in that it involves the *breakdown* of glucose) is temperature sensitive too. If, therefore, the greenhouse temperature is high when light levels are low, such as in a heated greenhouse in winter, the respiration rate may exceed the photosynthetic rate and valuable glucose may actually be wasted. There is a technical term – the **compensation point** – denoting the point at which the rate of photosynthesis (glucose gain) exactly balances the rate of respiration (glucose loss). A gardener should aim at maintaining rates of photosynthesis above the compensation point for as long as possible. However, without detailed information for individual plant species, and sophisticated monitoring equipment, it is difficult to know whether one is getting it exactly right. It's a matter of doing the best one can.

Gardening suggestion

Allow the maximum possible light into greenhouses and conservatories but try to regulate the temperature so that it does not get very hot - not always easy unless you have efficient thermostatically controlled air conditioning! Constant vigilance and opening and closing of windows are the only answer for ordinary conservatories and greenhouses. Thermostatically controlled ventilation (windows which open automatically as the temperature rises) is obtainable for greenhouses – not as expensive as full air conditioning. Blinds on both roof and windows are useful aids for emergencies.

Commercial growers sometimes boost the carbon dioxide supply in their greenhouses, but this is beyond the scope of ordinary gardening (see also the section on leaves in Chapter 2).

The global greenhouse effect – do trees help?

We hear a lot about the build-up of carbon dioxide in the air, thought to be contributing to the warming up of the planet. It is a popular idea that trees mop up carbon dioxide; therefore the more forests there are the better. It tends to be forgotten, however, that plants are respiring all the time, *giving out* carbon dioxide – just as we do when we breathe out. During the day this is masked by photosynthesis which involves the *using up* of carbon dioxide, but at night it is 'exhaled'. Scientists have actually measured the amounts of gas taken in and given off using a balloon-type apparatus. Mature trees – which are not growing – give out as much as they absorb, so are no help in combating the increase of carbon dioxide in the air. It is only actively growing, young trees which consume and store more than they give off. So planting a woodland of young trees will temporarily help to use up carbon dioxide. But when they are old and mature the benefit ceases, and no doubt the trees will either fall and decay (decomposition gives out carbon dioxide!) or you will burn them (ditto!). It's pretty much a 'no win' situation.

This argument can be extended to the pros and cons of real Christmas trees as opposed to plastic ones. Plastic uses up natural resources, but once made (unless it is the biodegradable kind) the carbon is 'locked up'. You could argue that a real Christmas tree is OK: you can replant it and it goes on growing indefinitely. But this is not a realistic proposition for more than a few years. Inevitably it will eventually be burnt or will decompose on the local authority's compost heap, releasing carbon dioxide back into the air. So, maybe, a plastic one isn't such a bad idea.

Making other molecules

Here is a brief summary of how plants proceed to make all the other things they need to grow and survive, using sugar from photosynthesis as their starting point.

- **Fats** (lipids), necessary for certain membrane structures in cells, are made from sugars because they, too, only contain carbon, hydrogen and oxygen.

- **Proteins**, however, need the addition of nitrogen and sulphur. Specialized proteins and vitamins, which may act as enzymes (molecules which are catalysts and help chemical reactions along), require such elements as magnesium, zinc and copper, and some others only in very small amounts.
- The **DNA** at the control centre of every cell needs phosphorus, and when plants are growing each new cell needs a copy of its DNA, which has to be made.
- Potassium and sodium are also essential elements in the functioning of cells.

So, when gardeners talk about 'feeding' their plants they are referring to the addition of these elements, which photosynthesis, on its own, doesn't supply (Table 1-1).

The term 'ion' refers to an electrically charged particle, which can move about freely in solution in water and can be absorbed into plant cells. The ions formed by atoms or molecules of metals – potassium, sodium, magnesium and so on – are positively charged and are called 'cations'. The gaseous or non-metallic elements form negatively charged anions, which are usually a group in which the element is combined with oxygen and sometimes hydrogen. For example, nitrogen forms nitrate (one atom of nitrogen combined with three of oxygen NO_3); phosphorus forms hydrogen phosphate (one atom of phosphorus

Table 1-1. 'Food' elements and the form in which they are absorbed (see also Table 4-2, Chapter 4, page 68).

Main elements (macro-nutrients)	*Absorbed as*
Nitrogen	nitrate & ammonium ions
Phosphorus	phosphate & orthophosphate ions
Potassium	potassium ions
Sulphur	sulphate ions
Magnesium	magnesium ions
Calcium	calcium ions

Trace elements (micro-nutrients)	*Absorbed as*
Iron	iron ions
Manganese	manganese ions
Copper	copper ions
Molybdenum	molybdenum ions
Zinc	zinc ions
Boron	boron ions
Chlorine	chloride ions
Nickel	nickel ions

Figure 1-4. Diagrammatic scheme showing how the main constituents of a plant are formed from the sugar initially manufactured by photosynthesis.

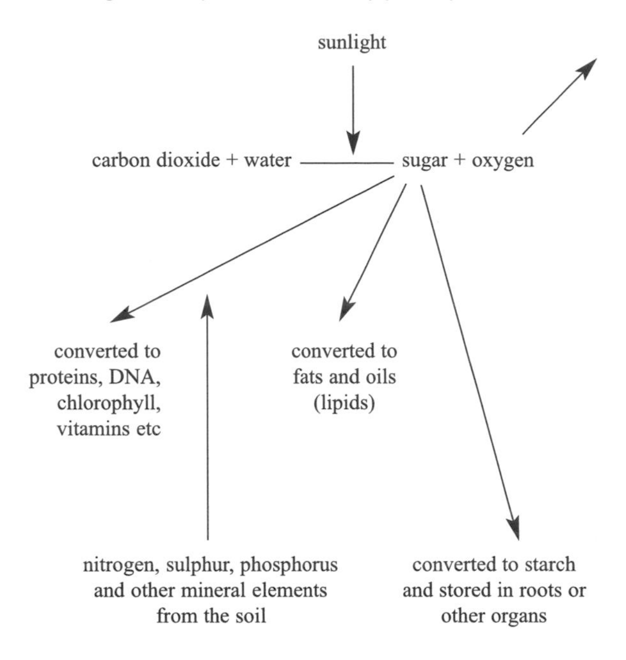

with four of oxygen and one of hydrogen HPO_4) and orthophosphate (one atom of phosphorus, four of oxygen and two of hydrogen H_2PO_4). In addition, there a few other elements needed by particular plants only – sodium (for the salt bush *Atriplex vesicaria*), cobalt for the nitrogen-fixing bacteria in leguminous plants, and silicon for certain grasses and for horsetails (*Equisetum* spp.).

The conversion of carbon dioxide and water into sugar, the storage of the sun's energy, and the subsequent manufacturing of the substances used to make cells are complex topics, but Figure 1-4 above should help.

The vital point here is that whether you are an 'organic' gardener, or one who uses the various chemical products from the shelves of a garden centre, the chemical substances involved in the processes inside a living plant are the same. 'Muck' is undoubtedly good for soil structure and water retention, and it makes sense to recycle organic material in the garden (see Chapter 4). But there is no thorough scientific evidence to suggest that there is anything special about manure and compost as far as the chemical elements they supply are concerned. Some types of compost are richer in certain elements than others, just as commercial tomato food, and a general fertilizer for non-flowering vegetables, differ in their constituents. As far as we know at present, there is nothing particularly mysterious about muck. Bacteria feed on it, break it down and release the constituent elements into the soil in mineral form so that they are available for plants to absorb in much the same way as they would absorb minerals from a man-made fertilizer (for gardening suggestions relating to nutrients see Chapter 4).

Water works

The special feature of life on this planet is that all the chemical reactions inside cells take place in a watery solution. Water is the mainstay of all that goes on in a plant – the reactions of respiration and photosynthesis all take place in solution in water and it is water pressure, acting in conjunction with the 'skeleton' of cell walls, that keeps a plant upright. Water has some remarkable and special properties, discussion of which is beyond the scope of this book; more advanced books are recommended in the Further Reading section. As gardeners we must be constantly alert to our plants' need for that precious liquid. More about how water is absorbed, transported and used will be explained in Chapter 2.

2. The parts of a plant – and looking after them

Scientists still have a great deal to learn about what goes on inside plants. Much research these days seems to be driven by political or commercial agendas, so it is very difficult to find funding for pure research into plant physiology (work which doesn't have any obvious practical or economic application). There is probably a lot we shall never know, but we can, with a certain amount of confidence, understand a good deal about their physical structure. This has been made possible by the use of microscopes and, more recently, the sophisticated electron microscope, which even gives images of structures inside cells. There are some life processes too, explained by the consistent results of many repeated experiments. This chapter takes a look at the bits of this knowledge which are relevant to gardening.

There is no such thing as a 'typical plant'. One of the things which makes gardening so absorbing is the amazing diversity of plants, and the intriguing ways in which they are adapted to different conditions. But most of those grown in our gardens sit with their roots in the soil and their stems emerging from the ground, holding the leafy bits and flowers up in the air. Aquatic plants have the same basic structure – it's just that the roots are anchored in the mud, or float freely in the water, while the stems and leaves are either submerged or peeping above the water surface.

Roots

Roots are very variable structures. Most are designed to anchor a plant in the ground, and to absorb water and nutrients from the solution between the soil particles. They may all spread out in a network near the surface, or there may be a large, central tap root from which smaller side-roots emerge. Large tap roots are often adapted for storage, for example, carrot, parsnip and swede. Some plants, such as tropical orchids, have roots dangling in the air, able to absorb water from the damp atmosphere, and many climbers, such as ivy and virginia creeper, have

Figure 2-1. Diagram to show how root hairs are tiny projections from single cells, each perhaps about 0.04 mm in diameter. They grow out between the soil particles and therefore increase the surface area over which absorption of water and minerals can take place.

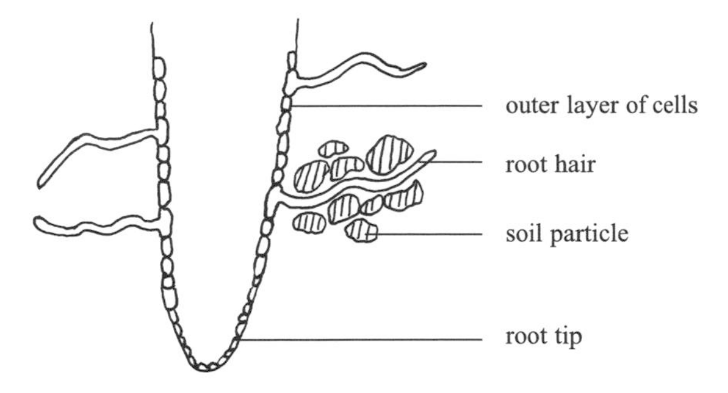

roots growing out along the length of their stems which are adapted for clinging to solid surfaces. What follows applies to roots which find themselves in soil.

Root hairs and absorption

There is literally much more to roots than meets the eye. With the aid of a microscope the region just behind the tip of a young root can be seen to be sprouting tiny projections from the surface cells. These **'root hairs'** are invisible to the naked eye, but they're all-important.

The root hairs enlarge the surface area through which water and mineral nutrients can be absorbed – and, although recent evidence shows that absorption can take place along the whole length of roots – the tips of young roots are where most of the action takes place. Ideally each root hair, and the surface cells further back along the root, should lie bathed in a watery solution of the minerals that the plant needs. If they are not bathed in water they shrivel and die.

This is where soil structure becomes relevant. Water clings to soil particles –

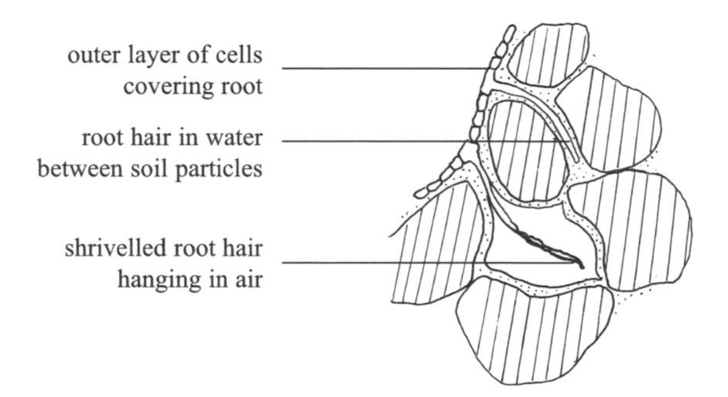

Figure 2-2. Diagram to show how in a sandy soil, where the particles are large and the spaces between them correspondingly extensive, a root hair might find itself hanging in an air space, rather than in a film of water. This is bad news for the plant.

round each grain of soil there is a film of watery solution. In water-retentive soils, such as clay, this can become more or less continuous and the problem then becomes lack of oxygen. Oxygen travels faster in air than when dissolved in water, so it's important that there are at least some air spaces connecting with the outside world to allow renewal of oxygen supplies (see also Chapter 4). But in sandy soils, where the particles are large, and the spaces in between wide, there is more danger of the roots finding themselves in an air space rather than in a film of water. Hence the need to be constantly aware of the state of the soil, to water adequately, and to plant only those species which are well adapted to coping with the particular type of soil (see Chapter 4).

Help from other friendly organisms

The other secret that roots hold is their close association with bacteria and fungi. Judging by the fungi-killing products on garden centre shelves, you could be forgiven for believing that all organisms closely related to moulds and mildews are harmful. Bacteria often cause disease and so are subconsciously blacklisted – but they are not all bad. Many plants depend on fungi to help them absorb nutrients from the soil, and on bacteria to supplement the supply of certain substances. It appears that evolution didn't get the design of roots completely right. Improvements, for the purpose of absorption, have been achieved by means of a little help from friends. These assistants are special fungi which, by virtue of their fine filamentous structure, are adept at creeping in between soil particles and ferreting out nutritious titbits. It has now been discovered that many herbaceous plants have these **mycorrhizae**, networks of fungi (singular: mycorrhiza) in and around their roots.

Although the spore-bearing parts of fungi, toadstools and mushrooms are more obvious and familiar, the main body of a fungus consists of fine threads which weave between soil particles or in amongst dead organic material or living cells. They may form a dense mat or mycelium. In many herbaceous plants these filaments, the **hyphae**, penetrate the roots and form a link between the root cells and the surrounding soil. They are able to absorb materials from the soil and transfer them to the tissues of the plant. Tree roots have similar associations, though it appears that here the fungal mycorrhizae form sheaths around the surface of the roots, rather than actually penetrating them. This is an example of symbiosis in which two partners live together in a mutually beneficial association. The fungus receives carbohydrates and vitamins from the plant and, since fungi feed on dead organic matter by exuding digestive enzymes to break it down into simpler soluble and absorbable substances, they can reciprocate by breaking down proteins in soil organic material to substances which can be absorbed and

Figure 2-3. Diagram to show how the thread-like structures (hyphae) of a fungus penetrate the tissues of a root, both between and actually inside cells. This illustrates how mycorrhizal fungi can help roots to absorb nutrients. But some fungi are harmful, and microscopic examination of a disease-causing fungus would look very much like this (see Chapter 5, page 91).

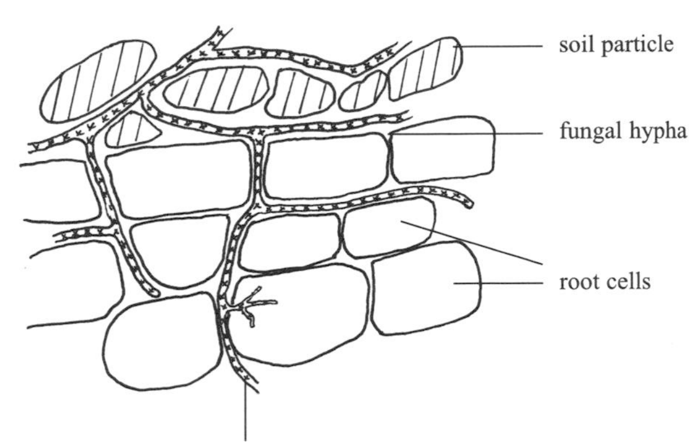

utilized by the plant. The network of fungal hyphae in the soil vastly increases the potential surface area for absorption over and above the root's own root hairs, so mycorrhizae are particularly important in poor soils where mineral nutrients are hard to come by.

Research has been throwing up ever more fascinating things about mycorrhizal fungi. For example, there is one group of plants, the cabbage family (Brassicaceae), which doesn't seem able to form this useful partnership. But orchids have gone a step further and have specialized arrangements in which the fungus can supply the plant with carbohydrate – very useful during the very slow developmental stages of the seedling. Indeed, certain orchids are unable to survive without their symbiotic fungus. Even more extraordinary, there is some evidence that individual plants can be connected by fungal threads – thereby being potentially able to feed each other! One genus of fungus, associated with the roots of alder trees, can (like nitrogen-fixing bacteria) utilize nitrogen from the air to manufacture compounds essential for making protein. The reason why certain toadstools are always found under particular trees is that they are the spore-forming bodies of the mycorrhizal co-habitant peculiar to that tree species. It has also been discovered that if the correct fungi are introduced into soil, together with new plantings on poor or badly disturbed sites, it can enhance the establishment of the new vegetation.

The specialized bacteria which can live inside root tissue and lend a hand there by fixing nitrogen are described in the section on the nitrogen cycle, Chapter 4, page 75.

Gardening suggestions

Watering in

Even if it's raining, it is a good idea to water thoroughly newly planted seedlings or plants transferred from elsewhere. This settles small soil particles round the roots and root hairs, and makes sure that they are properly in contact with a film of water, rather than stranded in an air space. It's a good idea to do this at some distance from the stem.

Organic matter

Manure or compost helps the roots. In sandy soils it fills in some of the air spaces, and provides more extensive films of water clinging round the particles of organic material. In clayey soils it can help to separate the tiny clay particles, and holds water whilst alleviating the lack of oxygen (see Chapter 4).

Planting and transplanting trees, shrubs and herbaceous perennials

If you leave roots exposed to the air the little root hairs and surface cells of the younger roots will soon shrivel up and be unable to absorb water and nutrients efficiently. Soak the roots of bare-rooted trees and shrubs in water, and cover them with a polythene bag. When transplanting try, if possible, to dig up a large root ball with soil intact. Don't shake off any soil; simply carry the lot, gently eased into a bag or pot if necessary, to the new site. In sandy soils this is tricky. If soil falls off, pop the roots immediately into a bucket of water for transference, and plant as soon as possible. The longer a plant is left out of its soil, the more the delicate surface cells will be damaged or will die, and absorptive capacity will be reduced.

Fertilizing (feeding)

If using a man-made fertilizer in solution (see also Chapter 4) be careful to follow instructions. If you apply too much, the solution in the soil may become too concentrated. If it is stronger than the solution inside the root cells the plant may become stressed through water loss (see osmosis, page 29). Try putting a piece of raw potato in a very concentrated salt solution; it will start losing water by osmosis and will begin to shrivel.

Being careful with fungicides

Sometimes it may be necessary to treat the upper parts of a plant with a proprietary product to get rid of a fungus infection. But remember: underground there may be fungi which are helping roots to absorb nutrients, so try to avoid spraying near the ground or spilling the fungicide on the soil.

Root training

Many plants like to spread fairly shallow roots, but there are those – root vegetables, for example – which like to develop a tap root and delve deep into the soil. Find out what the plants you are growing from seed like to do, and proceed accordingly. Special root-training modules can be obtained which encourage the roots to grow downwards rather than outwards and to develop a good strong main root. Or you can do it cheaply by using cardboard toilet paper rolls!

Planting container-grown or bare-rooted trees and shrubs

Container-grown trees and shrubs can be planted at any time of year, but bare-rooted ones need to be dealt with in winter when they are without leaves. In a container the roots are in a cosy root ball surrounded by soil particles which need not be disturbed during planting; the delicate younger roots – which do most of the absorbing – can carry on without interruption, and a continuous supply of water is carried up to the leaves. If, however, a leafy, bare-rooted tree is planted it will take some time for the young roots to settle in among the soil particles and to start to absorb again, during which time the leaves will be deprived of water, may wilt and may be in danger of not recovering. (See also leaves and transpiration, page 27.)

How deep to plant trees?

Most tree-root systems tend to grow out sideways rather than downwards into deeper soil. It is therefore thought better to spread the young roots of bare-rooted whips (small young saplings) in a fairly shallow hole rather than a deep one.

When to plant trees

There are two schools of thought – one is to plant in mid-autumn, the other to plant in early spring. The reasoning behind the autumn planting system is that the soil is still reasonably warm, and the roots more likely to grow and become established. The possible down-side to this is that the young roots then have to survive a cold, wet winter when fungus infections may set in. Advocates of planting in early spring, on the other hand, suggest that the roots will get off to a start as soon as the soil begins to warm up, and thereafter will grow strongly through the summer. Advice from Forest Research (an Agency of the Forestry Commission in the United Kingdom) is that so long as leaves have fallen before autumn planting, and the soil is neither frost-hardened nor excessively dry, both options are equally viable.

To test these hypotheses scientifically it would be necessary to do some large-scale trials with large numbers of identical young trees and two identical plots of land. One sample would be planted in mid-autumn, and one in early spring, and their growth measured and monitored over a period of perhaps two or three years. As far as I am aware no such trials have been carried out; it is difficult to find funding for such an experiment and it is tricky, too, to standardize the conditions.

Stems

A stem is a miracle of engineering. It is designed to keep the plant upright, the leaves exposed to the maximum light, and flowers visible to pollinating insects or exposed to the wind. At the same time it provides the route by which water can be carried up to the leaves, and manufactured food passed back down to the roots or to other cells which need it.

Support and transport – a scaffolding of pipes

These feats are achieved by means of specialized cells which become stretched out and joined together longitudinally to form microscopic tubes running all the way up and down the stem. The ones carrying water and dissolved minerals up are the **xylem** and the ones carrying sugars mainly down are the **phloem**. The hollow xylem tubes have lost both the cross walls between the cells and their living contents, and have walls specially strengthened by a very tough substance, **lignin**.

The 'woodiness' of trees and shrubs is due to particularly well-thickened xylem vessels, and lots of them. In trees and shrubs a cylinder of new xylem is laid down each year – hence the annual rings you see on the surface of a tree

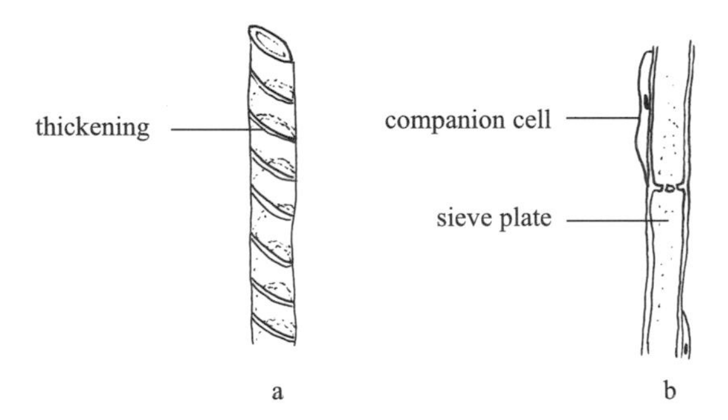

Figure 2-4. The structure of the two kinds of vessel in stems: (a) xylem, carrying water and dissolved minerals upwards; (b) phloem, carrying manufactured organic compounds, mainly downwards. The xylem vessel shown here has its wall thickened spirally, but there are other forms of thickening, the toughest ones being almost entirely reinforced, but scattered with 'pits' (thinner points in the thickening). These xylem tubes can be as much as several metres long and are formed by elongated cells arranging themselves end to end and the cytoplasm and cross walls disintegrating. Phloem vessels, on the other hand, have some living contents; they still have cross walls, the perforated sieve plates, and are accompanied by normal living cells, the companion cells.

stump, which can be used to estimate the tree's age. Figure 2-5 shows that, in a typical dicotyledonous plant, the groups of vessels are situated in a ring towards the outside. This has particular significance. Not only is it a good engineering principle to have your reinforcing rods arranged thus, it also enables the development of the extra thickening laid down year by year. The layer of dividing cells, the **cambium**, grows out sideways and joins up into a continuous cylinder throughout the stem. This dividing layer (a **meristem** – see Chapter 3) produces more phloem to the outside and xylem to the inside. The cause of light and dark rings on a tree stump is the differential growth of the vessels in summer and winter – in summer they tend to be larger and show up light in colour, in winter smaller, more closely packed and darker.

Plants that don't become woody (annual dicotyledons and all monocotyledons such as grasses and members of the lily family [Lileaceae]) have just enough thickening in their xylem to maintain the vessel as open tubes and to help to support the stem. In monocotyledons the vascular bundles, instead of being arranged in a ring round the stem, are scattered throughout. In addition, in most non-woody plants, there is some extra cellulose thickening in the walls of the cells around the edge of the stem, which is adjusted according to need. If the stems of plants growing along the edge of a busy road are examined microscopically, and compared with stems from the same species further away, extra thickening in these outer cells can be seen. The plant has responded to the stress of being swept around by vehicle slipstreams by strengthening its stem. This is particularly relevant when thinking about staking tall plants (see Gardening suggestions, page 32).

Plumbing and pumping

A system of pipes is all very well, but how is water raised from the soil up to the tips of stems and leaves? Imagine yourself as an engineer asked to lift water to the top of a 15-m high tree without a mechanical pump, let alone a source of power! How do plants do it? Scientific investigation has given us a rough idea, but there is still room for more understanding – particularly of the ability of sap to rise when there are as yet no leaves on a tree. When a tree is in leaf things are easier to explain. Leaves have a spongy internal structure – air spaces communicate with stomata (pores) which open to the outside. Water is constantly evaporating from the cells out into the air spaces and, from there, diffusing out through the pores. This is **transpiration** – not to be confused with respiration. (My husband and gardening assistant are always confusing them – I guess just to annoy me!) Go into a closed greenhouse on a warm day; the moisture transpired ends up as condensation on the glass.

Figure 2-5. Cross-section of a young dicotyledonous stem to show the arrangement of the different tissues and the different types of cell.
(a) Epidermis (skin) with a waxy covering, the cuticle.
(b) Collenchyma, a layer of cells just inside the epidermis which can be thickened with extra cellulose to strengthen the stem.
(c) Sclerenchyma, at the edge of each bundle of vessels and formed of very long, narrow cells with heavily thickened walls.
(d) Parenchyma, large, spherical cells with thin walls which make up the bulk of the body of the stem.
(e) Xylem vessels.
(f) Phloem vessels and companion cells.

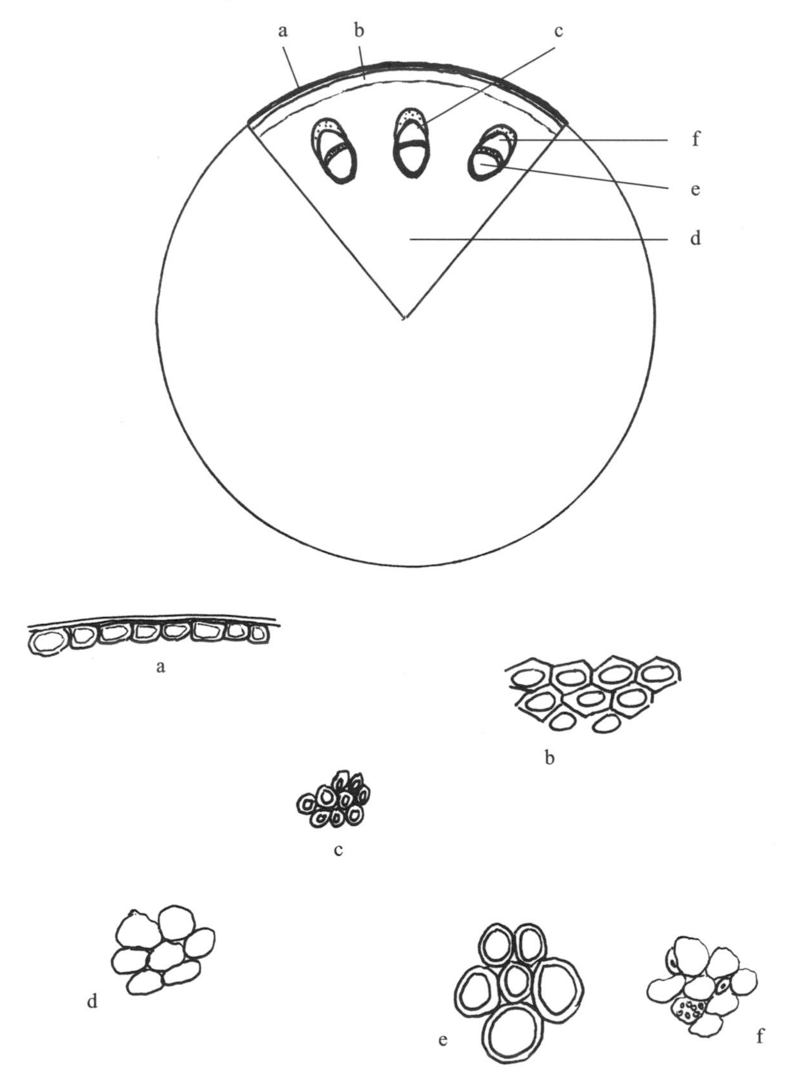

Figure 2-6. Diagram of stem structure to show the cells longitudinally (identification lettering as Figure 2-5).

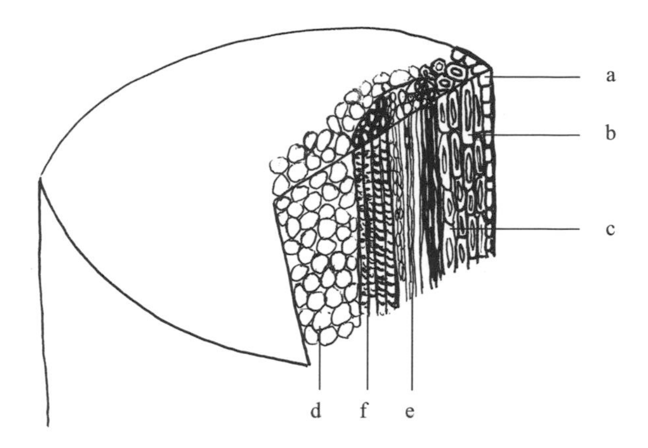

Transpiration creates a stream of water leaving the leaf cells, which in turn draws in water from neighbouring cells (see osmosis, below). Because of the cohesive properties of water molecules this stream eventually exerts a pull on columns of water in the stem's plumbing pipes. Water is literally pulled up the stem in this system of microscopic tubes. This, in turn, creates a suction effect in the roots, and encourages water to be absorbed from the soil, where there should be more of it than there is inside the roots.

If this isn't the case, or if the soil water is frozen, the plant is in trouble. Transpiration, being a passive physical process, goes on regardless. The leaves lose water but can't replace it from the soil and so, no longer turgidly full of water, the leaf and stem cells flop and the plant wilts (see below).

Turgid or flaccid cells – water stress

Every living cell is surrounded by a very thin 'skin' (membrane) which is permeable to water and to other small molecules, but not freely permeable to larger ones, therefore called **differentially permeable membranes**. Whenever a cell is in a situation where there are more water molecules outside it than inside – in other words situated next to a more dilute solution, which may be another cell or the solution in the soil – water will automatically pass into this cell. This movement of water is **osmosis** and happens all the time throughout the plant, be it in the root, where water is being absorbed from the soil, or in the leaves, where water is being lost by evaporation. Water will pass into a plant cell until the pressure of the swelling cytoplasm, pushing outwards, is counteracted by the inward pressure of the relatively rigid cell wall, and the cell can expand no longer –

imagine a balloon blown up to the point of bursting. In this condition a cell is said to be **fully turgid**, and this is how healthy plant cells should be most of the time. Non-woody plants stay firmly upright, and leaves stay spread out, because of this turgidity – imagine a plant's tissues as stacks of fully blown-up balloons. As soon as cells fail to obtain enough water to stay turgid (and so become floppy or flaccid) that part of the plant is in danger of wilting.

Drought conditions in the soil soon put a plant in jeopardy because, as explained above, transpiration goes on regardless. The leaf and stem cells lose water, and wilting will soon occur unless something is done to delay it. Miraculously there is a mechanism which comes into play as soon as the water supply to the roots is reduced. There is evidence that the root tips send a signal to the leaves, probably in the form of a hormone (see Chapter 3) which tells the leaf pores (stomata) to close. Transpiration is thus reduced – though not entirely stopped – and wilting is delayed. The situation for the plant, however, is not ideal and it is said to be under **water stress**. In this condition many chemical processes in the cells don't function normally, and growth tends to be inhibited. The only way of relieving this, and preventing the plant from eventually wilting and finally dying, is to replenish the water supply. If the plant has been stressed for some time recovery may be slow.

The fact that transpiration carries on regardless – unless there is some physical action to stop it – may help to explain why deciduous trees and herbaceous perennials lose their leaves in winter. If the soil water freezes, as it may do in temperate latitudes, the situation is similar to severe drought – water being no longer available in liquid form. Leaf fall, and the sealing off of the scars, stops transpiration and the plant can conserve water until the spring. Evergreens, cacti and some other specialized plants are adapted for avoiding water loss. They employ strategies such as reducing the surface area of their leaves until they are as thin as (pine) needles; having a very thick waxy cuticle; and/or stomata positioned to give protection from the wind (see leaves, page 35). Such plants are known as **xerophytes**.

Gardening suggestions

Handling seedlings

When handling tiny seedlings, think of their delicate and newly formed plumbing (xylem and phloem vessels) – they are easily damaged. Pick them up by their seed leaves rather than their stems. They can always grow new leaves but can't easily replace damaged vessels – at least not in time for the essential water and nutrients to

Figure 2-7. The kindest way to handle a seedling.

be transported to where they are needed. If there is a good root ball then gently support it from underneath; a teaspoon is useful.

Lifting shrubs and trees

Support the plant from underneath, by slipping one hand or a piece of polythene or sacking under the roots. Alternatively, transport the specimen in a wheelbarrow, after supporting it below for the lift-up. How often have you seen those TV 'make-over' gardeners lugging shrubs around by their stems? If you just hold the stem or trunk, the heavy root ball will be pulling down and exerting an unnecessary force on the internal tissues: imagine being lifted up by your neck!

Staking

Provide support for tall herbaceous plants before they actually need it. If tall plants are allowed to wave about in the wind they will waste energy by adding extra thickening to cells in their stems (see page 27). You want them to conserve their energy and resources for growing beautiful foliage or flowers.

Keep an eye on ties to make sure that they aren't too tight, either cutting in to the delicate conducting vessels or restricting them.

Anticipating water stress

Even before wilting is apparent, a plant which is not receiving enough water is neither happy nor healthy. As soon as the weather forecasters predict a period of dry weather, get the hose ready and don't wait too long before watering those plants which you know from experience, or from growing instructions, are not well adapted to dry conditions. Water straight down at the base of vulnerable plants, rather than using a sprinkler. Keep a wary look out for the slightest sign of wilting leaves and get to know the ones at risk. Examples in my own garden are phlox and rudbeckia. A moisture metre, available from garden centres, is very useful for checking plant pots indoors.

Leaves

Leaves are the plant's solar panels. Hence they are usually flat and thin and, in most plants, spread out at right angles to the sun's rays. Exceptions to this are the monocotyledonous plants with parallel-veined leaves – the lily family and grasses, rushes and sedges. Take a look, too, at how leaves are arranged on the stem; there is minimum overlapping so that each leaf gets as much light as possible.

Capturing light and manufacturing food

A typical dicotyledonous leaf has its main food-manufacturing layer near the top where it gets most light (for photosynthesis see Chapter 1, page 14). Below this layer is a spongy region with air spaces communicating with the outside through the stomata. These pores tend to be more numerous on the under surface – a way of preventing too much water loss on sunny days when the upper surface heats up. They are also cleverly designed with a pair of specialized cells on either side of the opening. These so-called **guard cells** can change shape and close or open the pore according to need – they are open during the day but close at night, or at times of water shortage. The leaf surface is protected by a skin or cuticle of waxy material – yet another miracle of engineering!

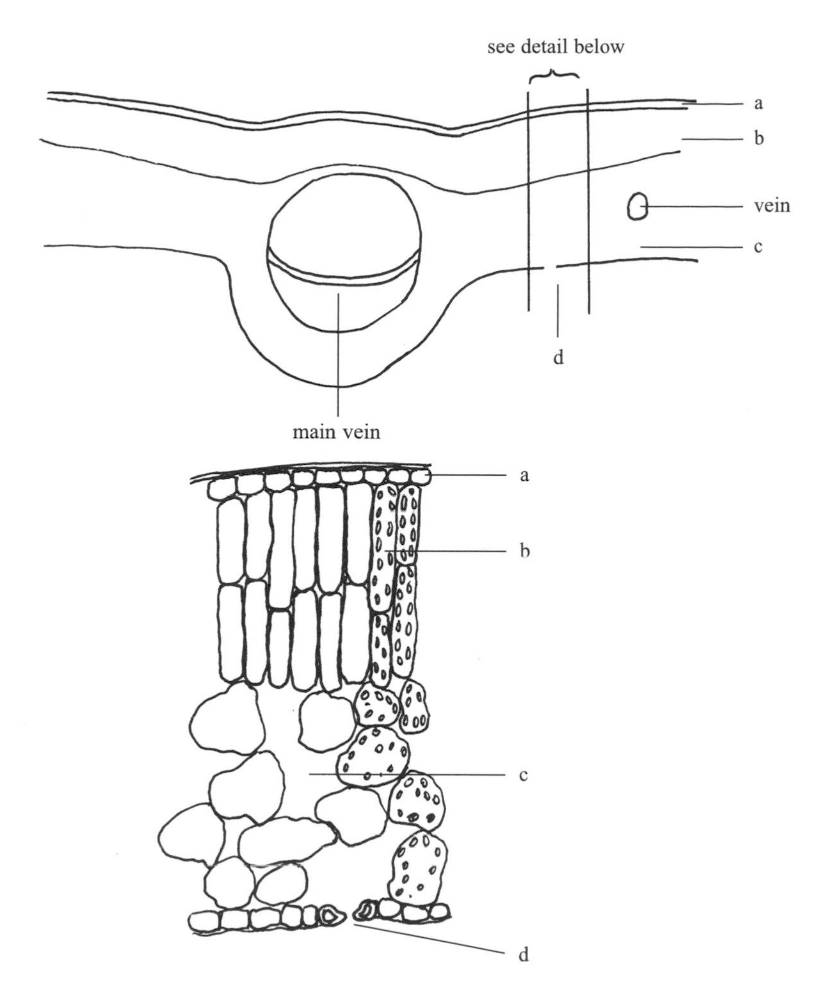

Figure 2-8. Section through a typical leaf of a broad-leaved dicotyledonous species to show the arrangement of the tissues and the different kinds of cell.
(a) Epidermis with a waxy covering of cuticle.
(b) Palisade layer, the main site of photosynthesis: the elongated shape of the cells at right angles to the surface probably works as an adaptation to allow the maximum amount of light through to the chloroplasts (if arranged the other way round there would be more cell walls to penetrate).
(c) Spongy mesophyll layer, with many communicating air spaces for the exchange of carbon dioxide and oxygen.
(d) Stoma (plural stomata), surrounded by two guard cells which can open and close the pore. In plants which hold their leaves at right angles to the sun's rays most of the stomata are on the cooler lower surface – an adaptation to prevent undue water loss. There are many variations – for example, in monocotyledons (grasses and so on, in which the leaves tend to be held vertically) there are equal numbers of pores on both surfaces. In conifers the leaves may be needle-shaped to reduce the proportion of surface area, and in specialized xerophytes there are many weird and wonderful adaptations to prevent water loss – curled shape, extra thick cuticle, hairs, sunken stomata, and so on.

Gardening suggestions

How much light and where to plant

The light requirements of different species vary greatly, the result of millions of years of evolution in different habitats. All our garden plants were originally bred from wild plants, some of which were adapted to woodland conditions, some to open meadowland, some tropical, some temperate, and so on. Those which originated in shady forests need relatively little light, those from woodland edge habitats a bit more – but not direct sunlight all day – whilst grassland species require full sun. There are numerous books that provide information about this, and plants bought at garden centres have labels with planting instructions. The best policy is to abide by the suggestions. If you try to put a plant in a place to which it is not adapted you can't expect it to flourish.

Keeping leaf surfaces clean

It is obvious that efficiency is going to be reduced by anything that stops light getting through to the food factory, or is inclined to block up pores. Try to keep leaf surfaces scrupulously free of debris. When planting out, and watering in, it is very easy to leave grains of soil on the leaves. It's particularly important for seedlings to have clean leaves, so wash off any soil particles either by using a small watering can with a gentle spray, or by just sprinkling a few final splashes with your hand. Indoor or conservatory plants should be regularly dusted or gently wiped with a damp cloth. Mildew on leaves is also going to reduce absorption of light, so try to control it, either with a fungicide spray or by careful hygiene.

Keeping spring bulbs going

A bulb is a food store, designed to supply enough reserves for the following year's growth. So if you want your daffodils, tulips, crocuses and so on to do well every year, it is essential to let them stock up well for the future. This involves allowing the leaves to carry on photosynthesizing and sending sugars down to the underground bulb or corm, where they can be converted into other substances for storage through the late summer, autumn and winter. Perhaps apply some nutrients (see Chapter 4), and don't cut down the leaves until they start to turn brown. They may look a bit untidy, but it's well worth it.

Tree pruning

Leaves tend to close their pores under very sunny, hot conditions – a protection against losing too much water (see transpiration, page 27). The exposed outer branches of a tree are most likely to suffer, and will therefore probably have the greatest proportion of closed stomata. Meanwhile, smaller inner branches in the crown are doing most of the work of taking in carbon dioxide for photosynthesis and keeping the transpiration stream sucking water up from below. When pruning, therefore, it's a good idea to leave plenty of protected small branches - concentrate rather on taking out large old-established branches.

Plants with leaves adapted to reducing water loss

If your garden is prone to drought you may need to look out for plants which can tolerate dry conditions. Their technical name is **xerophytes**. Characteristics of such plants are leaves with an especially thick, waxy covering, or with a thick covering of hairs; there are also leaves which are fleshy for storing water, or reduced to a narrow needle-like shape, which cuts down their surface area in proportion to their volume and so lessens the surface from which water can evaporate. Examples of herbaceous plants are *Mesembryanthemum* spp. and *Sedum* spp. (fleshy leaves), broom (*Cytisus*), juniper (*Juniperus*) and herbs such as lavender (*Lavendula*) and rosemary (*Rosmarinus*) with narrow needle-like leaves, and many alpines (hairy leaves). Conifers with needle-like leaves are the obvious examples amongst trees. There are many more examples, and lists in gardening encyclopaedias (see Further Reading list).

It's worth thinking about the natural habitats of some of these drought-resistant plants. Mesembryanthemums abound in the wonderful semi-deserts of South Africa's Western and Northern Cape regions. You find broom and gorse on dry heathlands throughout Europe, and lavender and rosemary are typical of the Mediterranean garrigue - the scrubby, arid terrain found in southern France, Italy and Greece.

Flowers and seeds

Plants have the edge on us because, if they don't feel like sex, many of them can do without it and still propagate themselves (see Chapter 3, page 56).

Sex lives

As a general rule, though, flowers are for reproducing, and are designed to operate the time-honoured system – male getting together with female. Those plants which have flowers belong to a group formerly called the Angiosperms which literally means 'covered seeds' – arising from ovules protected in an ovary. Terminology changes, though, and the group is now called the **Magnoliophyta** (see Chapter 8).

Figure 2-9. Typical flowers, showing the basic structure with different positions of the ovary.
(a) Ovary 'superior' (above the origin of the petals and sepals) and made up of several separate carpels, each with a single ovule.
(b) Ovary 'superior' and made up of several carpels joined together, each compartment containing several ovules.
(c) Ovary 'inferior', that is, below the origin of the other flower parts but, like (b), made up of joined carpels.

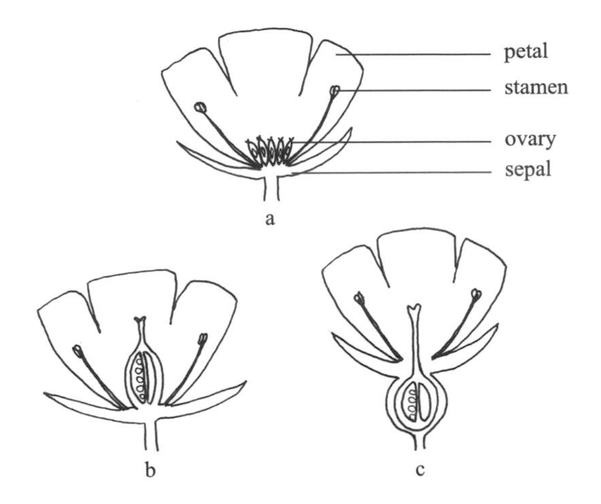

Pollination

First of all, male pollen from stamens has to be transferred to the female stigma. This may be mediated by insects or by wind and may be self pollination (transference within the same flower or to a flower on the same plant), or cross pollination (transference of pollen from one plant to another). In the wild both strategies may have advantages.

Self pollination (in effect, inbreeding) gives a rather uniform population but, provided that the environment doesn't change, can result in lots of well-adapted individuals. It is practised by groundsel (*Senecio* spp.) and chickweed (*Stellaria* spp.), both successful plants, as gardeners know only too well. The disadvantage of continued inbreeding is that plants can suffer loss of vigour – a condition in horticultural breeding programmes called **'inbreeding depression'**.

Cross pollination, on the other hand, results in more mixing up of genes to give greater variation and, in addition, hybrid vigour of heterozygous individuals (see Chapter 7), so there is more chance of flexibility in adapting to a changing environment. Cross pollination is the norm in the wild and there are all manner of arrangements for making sure that it occurs, from attracting insects with nectar and pollen, to bearing male and female flowers on different plants, or having anthers and stigmas mature at different times. Both systems exist in cultivated plants though, in general, perennials can't be easily self pollinated.

Fertilization

In animals one can understand that a male sperm cell, which is usually a swimmer of Olympic stature, can be attracted to a female ovule by some chemical signal and will move towards it quite easily through a liquid medium. But how is the microscopic male nucleus in a pollen grain on top of the stigma going to reach the female gamete in the ovary, which may be as much as a centimetre away down the solid style? It is a matter of quite extraordinary chemistry, concerning concentrations of certain chemical substances and their effect on the proteins in the tip of a tiny root-like structure, the pollen tube. This sprouts out of the pollen grain and finds its way down the style to the ovary and into the ovule via a tiny pore. At the tip of the pollen tube is a nucleus – the 'tube nucleus' – that directs the growth process. Behind this are two other male nuclei, one destined to fuse with the female nucleus in the ovum to form the embryo which eventually grows into the new plant, and the second which fuses with another female nucleus to form an 'endosperm nucleus', later concerned with the seed's food store. It's a lot more complicated than fertilization and embryo formation in animals. Plants may

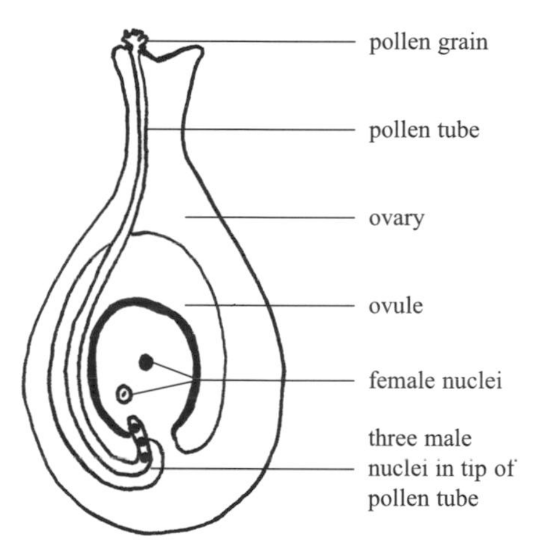

Figure 2-10. Fertilization. In real life the situation is a good deal more complicated, with several other cells and their nuclei involved.
The essential features (shown here) are the growth of the pollen tube with its three nuclei; the important female nucleus (which, when fertilized by one of the male nuclei, becomes the embryonic new plant); and the other female nucleus, to be fertilized by the other male nucleus, which helps to form the food store of the seed.
The fertilized ovum, together with surrounding layers of ovule tissue, then becomes a seed inside the old ovary. What was once the ovary becomes the seed case, fruit or vegetable. Sometimes some other structure swells to form the fruit (see Figure 2-11). A seed is arguably one of nature's most miraculous inventions - all the materials and instructions necessary for creating a new plant are packaged in a tiny, apparently lifeless little granule. (For more about seeds see Chapter 3; for gametes see Chapter 7.)

Figure 2-11. Diagrams on longitudinal section to show different kinds of fruit formation.
(a) Plum, the edible, fleshy bit being part of the swollen ovary wall, the ovary being a single carpel with a single seed (a blackberry is a collection of very similar structures).
(b) Pea pod, in which a single carpel contains several seeds (cut a tomato in half and you'll see two, three or more carpels, each with many seeds).
(c) Strawberry, in which the fleshy part is actually the swollen upper part of the stem on which are situated numerous single carpels, each one containing a single tiny seed.

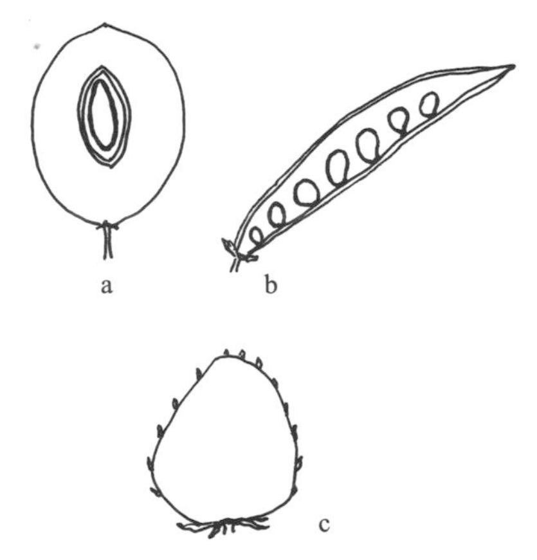

never have brains, but they achieve some remarkable feats! Figure 2.10 shows what happens in simplified outline.

Flowers for decoration or productivity

The function of flowers is to produce seeds. But if you simply want colourful flowers and are not interested in collecting their seeds, it is perhaps a mistake to let the plants waste energy on producing them. If, however, you want the end product as a vegetable, for a source of seed for next year, or simply to provide food for birds, then fruits and seeds have to be nurtured appropriately.

Flowers without colourful petals

There are many flowering plants which don't have conspicuous coloured petals – the grasses, many shrubs and most trees are examples. These are adapted for wind pollination and produce masses of very fine pollen which can be wafted quite long distances on the wind. Some of the tree species are **dioecious**, which means that there are separate male and female plants. My neighbour once came to me, quite mystified, to show me a twig from a bush in her garden which had catkin-like structures but without any sign of stamens or pollen: the female flowers from the female bush of a pussy willow.

Examine, with a good magnifying glass, the flower of a grass or a tree, and the equivalent of petals and sepals can be seen, even if they are only scale-like. There are also recognizable stamens and carpels containing ovules.

Gardening suggestions

Encouraging pollinating insects

If you want seeds, fruits or vegetables to develop, pollination will be essential. Bees are a gardener's best friend for this, and there are ways in which you can encourage them to visit: provide plenty of plants with nectar, and provide nesting sites for bumble bees (see Chapter 6, page 126).

Hand pollinating

This may be necessary in a greenhouse or conservatory. To pollinate tomato flowers, try using a piece of fluffy material attached to a stick - a pollinating 'mop'. Gently tickle the flowers with the fluffy mop.

Dead-heading

If you are aiming at extending flowering as long as possible, and are not interested in seeds or fruits, cut off seed heads in their prime. The plant then doesn't waste energy which it could use for producing more flowers. It's important to remove the whole ovary. Just pulling off the petals is of no benefit - very easy to do inadvertently on certain plants, for example, fuchsias, petunias and busy lizzies.

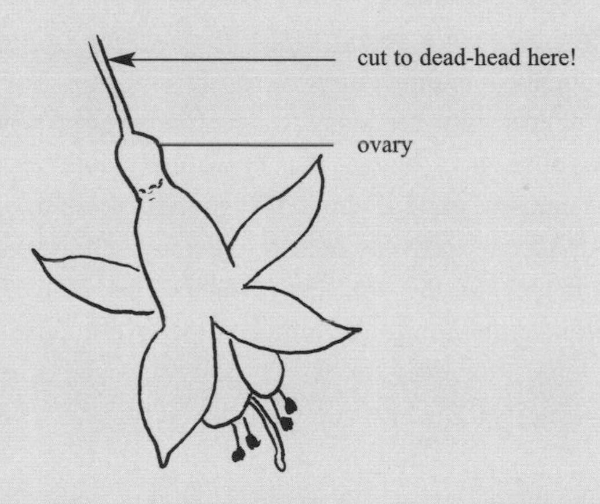

Figure 2-12. Where to dead-head.

Watering for good fruit development

Vegetables and fruits which are derived from the development of the whole ovary (for example, beans and tomatoes) or even the swollen upper part of the stem (such as strawberries) are made up of lots of large water-filled cells. It's therefore vital to give them copious water supplies.

Keeping seeds

If you want to collect and keep seed, remember that a mature seed is protected by a thick outer coat, and it remains viable and dormant only if kept dry. Wait until seed heads are completely brown and dried out, then shake the seed into a paper bag or envelope (not polythene as it tends to collect condensation and create damp conditions which attract mould). Keep them in a cool, dry place. Store fairly large seeds spread out on sheets of kitchen paper placed in layers in a box or drawer. The whole sheet, with seeds intact, can then be planted directly into compost.

Scattering seed before removing annual

If, during mid-summer, you are having a clearance of annuals such as forget-me-nots or poached-egg plant, give the seed heads a good shake over the ground before consigning the rest to be composted. This is an effortless way to ensure a good display again the following year.

Why do flowers have such an amazing variety of shape and colour?

Think back 225 million years, to the Triassic era, when fossil evidence shows that dinosaurs were beginning to evolve and there were probably no colourful flowering plants. How they came to exist is still something of a mystery. Scientists rely on fossils for their interpretation of the distant past, but the problem with flowers is their delicacy and tendency to decompose quickly which makes them less likely to fossilize. It is thought that flowering plants began to evolve some time during the Jurassic period, about 150 million years ago, and that the first ones were woody species, possibly related to the magnolias (magnolia-like pollen has been found in rocks of this era). However they originated, the need to attract creatures to pollinate them must have been a major influence in their evolution. Thus bright colours – or even, as we've now discovered, patches or patterns reflecting ultra-violet light, which can be detected by bees – would have been an advantage for attracting passing insects. Scent would have been important too. Bees seem to like the odours which we also like, but many flies are attracted by smells which we find obnoxious (flowers pollinated by flies are not ones to give to your loved one!).

Shapes, too, matter. In an open grassland habitat a radially symmetrical shape (star-shaped, daisy-like), held facing upwards would, perhaps, be the best way of attracting a creature flying directly overhead. But the situation is different in a woodland in summer, where flowers would be seen against a background of trees in leaf. So bell shapes arranged in a row down one side of a stem away from the

Figure 2-13. Two basic flower shapes: (a) star-shaped, radially symmetrical (actinomorphic) flowers tend to be held with their faces upwards; (b) bilaterally symmetric (zygomorphic) ones tend to be held facing sideways.

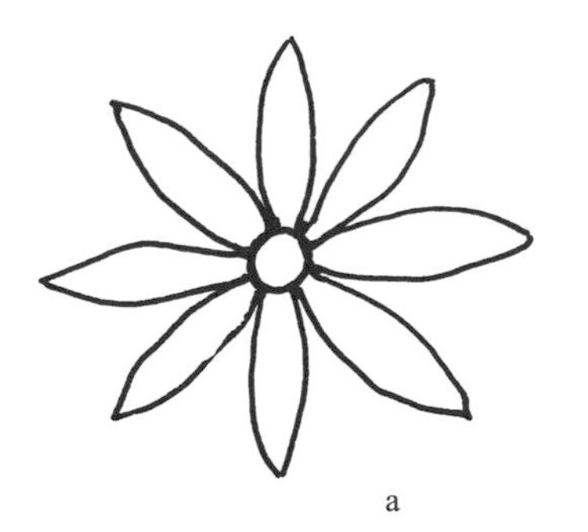

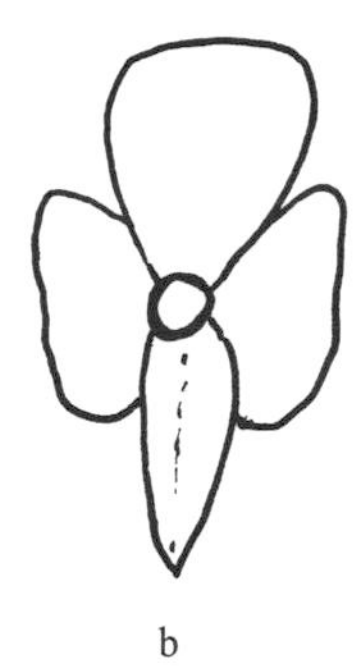

trees, such as foxglove or bluebell, or a sweet-pea-type shape (bilaterally symmetrical) with a 'landing platform' facing outwards, such as vetch, would be noticed more easily. There are star-shaped woodland flowers – wood sorrel, snowdrop, our garden hellebores – but they are invariably held drooping downwards rather than with their faces upwards. The exceptions are lesser celandine and wood anemone which hold their star-shaped faces upwards, but these two, and the three mentioned previously, flower very early in the year, before there is a canopy of leaves – in other words when the habitat is open and is not yet a true woodland.

Timing of flowering

A great deal of research has been done on this topic, mostly relevant to the horticultural industry and so not covered here. There are suggestions for further reading at the end of the book.

Non-flowering plant

Visit a horticultural show, and the immediate impact is that of the blaze of colour boasted by the flowering plants. But conifers and ferns are not unimportant denizens of gardens, while the mosses, liverworts and algae are ever present despite their occasional undesirability. The crucial difference is that the flowering plants have evolved a much more sophisticated method of reproduction compared with their more lowly relatives. The conifers (once included in a group called the Gymnosperms – 'naked seeds', but now having a Division of their own, the Coniferophyta; see Chapter 8) – have ovules situated at the bases of the scales of the female cones, and not enclosed in a protective ovary as in the flowering plants. The ferns have a complicated life cycle, with two different stages – the gamete-bearing, almost microscopic form, from which grows the spore-bearing adult plant that we know as a fern. Suffice it to say that the process of fertilization involves motile male gametes, which have to swim in water to find the female cells, and this is the case, too, for the mosses and liverworts. The algae mostly live in water and, in freshwater, take the form either of single cells or colonies of cells in spheres or filaments. The green slime which plagues garden ponds is usually a filamentous member of the algae. They have weird and wonderful methods of reproduction, sometimes involving sex between filaments.

Chapter 3 looks at seeds and how the amazing feats of germination and development are achieved.

3. From seed to senescence

Seeds and germination

A seed is one of nature's most supreme miracles. It is *the* DIY, flat-pack kit *par excellence*, complete with the full set of instructions for creating a grown-up plant. All the gardener has to do is to provide the right conditions to set development in motion and keep it going. My husband proudly boasts of producing over 650 grape tomatoes on one plant, grown from a single tiny seed saved from the previous year. (But, surely, the seed should claim the credit!)

If a thin section of a seed is cut and examined microscopically three important basic features stand out:

- A small embryo in one corner, the potential new plant.
- The food store, occupying the rest of the seed.
- The thick protective coat that surrounds the above.

The **food store** is an important source of organic compounds which provide the growing embryo with all it needs before it can begin to make its own food by photosynthesis. The **seed coat** (testa) is vital for protection during the period of dispersal and possible dormancy, when the seed may be subject to all kinds of unfavourable conditions. There are many variations on the basic theme – some seeds are very tiny, some enormous, some are unusual shapes, many have special structures to aid dispersal and some have particular problems. Examples of the latter are the seeds of many orchids. These are so tiny that there is no room for much in the way of a food store, and they have to rely on help from friendly fungi to supply food during germination and early growth (mycorrhizal fungi – see Chapter 2, page 22).

Amongst the flowering plants there are two distinct kinds of seeds – dicotyledonous and monocotyledonous. The 'dicot' type has two leaf-like structures, the **cotyledons**. These sometimes appear as the first two green leaves after germination, and are often called the 'seed leaves'. In some species the cotyledons swell up and act as the main food store while the original food store shrivels up; for example, in peas and beans the part that we eat and find nutritious is formed by the cotyledons. The possession of two of these cotyledons has

Figure 3-1. Basic seed structure: (a) section through a dicotyledonous seed, showing the two cotyledons; (b) section through a monocotyledonous seed, in which the single cotyledon has shrivelled away and has been replaced by a large food store, the endosperm.

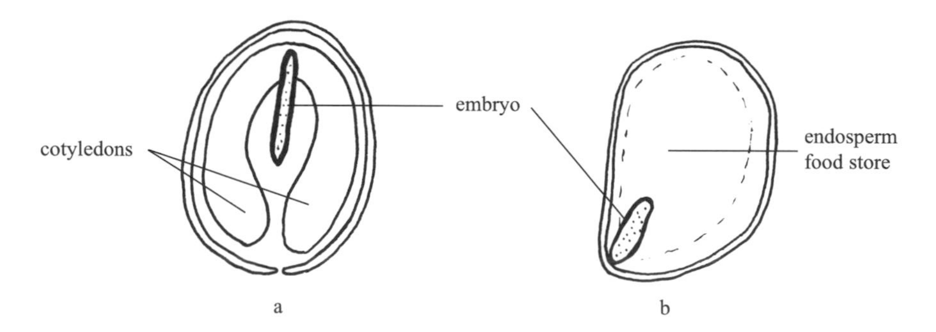

given rise to the botanical term for this big group of plants (**Dicotyledones**), their other characteristics being leaves with branching veins and a great variety of shapes. Their flower parts tend to be in fours or fives rather than in multiples of three. The 'monocots' (**Monocotyledones**) have a single cotyledon, and the mature plants differ from the 'dicots' in having simple, parallel-veined leaves and flower parts usually in multiples of three. The cereal species which we have cultivated are 'monocots', the nutritious part for us being the original food store, the **endosperm**.

Germination

It is stating the obvious to say that water and warmth are the wake-up call for a seed. But what actually happens? The story starts in the food store, where there are specialist enzymes (agents which enable all the chemical reactions in a living organism to take place in a safe, controlled way, at reasonable temperatures). Enzymes, though, only work in watery solutions and within a certain temperature range. Thus dry, cool seeds will remain dormant, but as soon as **water** creeps in through the tiny pore in the seed coat, and then through cracks formed by the swelling contents, the enzymes dissolve in water. If the temperature then reaches its optimum they start working.

The next requirement is **energy**, so the enzymes concerned with converting stored starch to glucose get going. Then enzymes dealing with respiration swing into action (see Chapter 1). Provided there is plenty of oxygen, energy is made available for **growth** to begin. Stored proteins and fats are broken down into soluble smaller molecules, which are sent off to the embryo. Here they are assembled into the components needed to make new cells – a tiny root forms at one end, and a shoot at the other. These eventually burst out through the weak-

Figure 3-2. Illustrating (a) epigeal and (b) hypogeal germination.

ened seed coat. From the seeds of most garden plants the food store also emerges in the form of two seed leaves or cotyledons. These are green and can do some initial photosynthesizing until new leaves appear. But in others, the monocotyledons (grasses and cereals), the food store remains behind and eventually shrivels away.

There are two ways in which the young plants emerge from the seed – but the fascinating thing is that the shoot always curls over as it pushes through the soil so that the delicate growing point is protected, and the strain is taken by the curved edge of the stem. In so-called **epigeal germination** the young stem curls over, and pulls up with it the cotyledons, often still surrounded by the seed coat. The cotyledons then act as photosynthetic organs until the young shoot appears between them. If you've grown sunflowers or pot marigolds from seed you will have noticed some of the seedlings with the seed coat still attached. In **hypogeal germination** the seed coat and cotyledons remain in the ground and eventually rot away while the young shoot pushes its way up through the soil with the young stem curved to protect the growing point. Beans do it this way. Next time you have some seeds germinating see if you can detect which method they employ.

Seed dormancy

However, some seeds don't automatically germinate when watered and warmed up. Some plants (including, not surprisingly, many alpines) need a period of cold before they will germinate (see Stratifying, page 48). Others, strangely perhaps, need the stimulus of light after absorbing water. An example is lettuce, some varieties of which will only germinate if exposed to light. Many weed species also need light for germination (bad news if some have got buried and you dig them up in the spring!). This is probably an adaptation to avoiding growing up in a shady situation which they can't tolerate. The control of light-induced germination is mediated by a chemical called **phytochrome** (see page 55).

When and how to sow

The optimum time and temperature for germination for any one plant depends a lot on its provenance; if it originated from tropical or Mediterranean climes the temperature required may be higher than that for a plant originating from a temperate or alpine habitat. Again, there is always plenty of information on seed packets. Remember that the seed needs water to start the chemical reactions, but it also needs oxygen so the compost must not be allowed to get soggy and waterlogged; too much water between the soil particles prevents air getting to the seed.

Gardening suggestions

Inside or outside?

If in doubt - inside. It's far better to have some control over the conditions for germination so, whatever you are growing, sow in pots or trays under glass if you want to be certain of success. If you don't have a greenhouse there are many kinds of propagating devices on the market from mini-greenhouses to heated trays. Instructions on packets relating to timing and temperature are the result of many generations of research and experience, so follow them as precisely as possible!

To soak or not to soak?

Common sense tells you that a big seed might have more problem getting water to all parts of its interior than a small one. It is a good idea to immerse really big seeds, such as peas (including sweet peas) and beans, in water for up to 12 hours. But remember the need for oxygen - any longer and your seeds may die for lack of air.

What medium to use?

If you want to save money, and have a good loamy soil in the garden, there is nothing against using that - provided you sieve it to make a fine seed bed before putting it in pots or trays. You could even do as was done in the days before commercially prepared composts - sterilize it with boiling water and drain well! The advantages of specially prepared growing media (not to be confused with the compost you may make on your garden heap, which is for mulching and improving the soil) are that they are sterile. They also have a good balance of nutrients, with some lime to give a suitable pH (see Chapter 4, page 79). John Innes 1, 2 and 3 refer simply to recipes designed for small, medium and large plants, all containing a mixture of loam, peat and sharp sand. No. 4 is lime free and is for acid-loving plants. There are loamless composts made from a variety of dead plant material (peat, coconut waste and so on) but, if you neglect watering for a while, they are difficult to re-wet. To be completely peat free (see Chapter 6) you must choose one of the media based on recycled plant material. These are improving all the time – look out for the latest versions.

Keeping damp but avoiding 'damping off'

Moisture, but not too much, is the golden rule – if possible stand your container in a tray and water from below (this also prevents seeds being dislodged). If you can use tepid water, rather than giving your seedlings the shock of a cold shower, all the better. Covering with cling film helps to retain moisture and reduce the need for constant vigilance. 'Damping off' is a fungus disease of the stem which may set in if the conditions are too wet and warm. The stem then collapses – watch out for this and keep the seedlings cooler and drier if necessary.

Temperature

There is tremendous variation in the temperatures at which seeds will germinate: it all depends on where the plants originate. Plants from arctic or alpine regions, perhaps surprisingly, have seeds requiring high temperatures – this is an adaptation to making sure that germination takes place at the height of summer when there is no danger of seedlings getting frost-bitten. Tropical plants, too, need high temperatures, but Mediterranean plants tend to sprout when it is quite cold; the best time for their seedlings is autumn, when the summer droughts are over and rain comes. So follow the instructions on the packet – the knowledge gained by experience and knowledge of provenance!

How deep to sow?

Look at your seeds and imagine how robust or delicate the shoot is going to be. Very small seeds will have very tiny shoots, and if planted too deep will find pushing up to the surface too hard. Larger seeds can be sown deeper. A good general rule is to cover the seed to a depth of its diameter. Very fine seed can be simply left on the surface.

Even distribution

This is important to avoid competition for light and nutrients between seedlings which are clumped close together. Develop your own technique – but don't just pour straight from the packet! A good system is to place small quantities at a time on the palm of the hand, and use a gentle tapping motion with the other. Another idea is to use an envelope with a tiny hole in one corner, or even to mix your seed into a flour-and-water paste and use an icing bag (but how many people these days go in for icing by hand?).

Stratifying

This is the treatment some seeds need in order to persuade them that they have actually been through the winter and it is time to come to life. Certain plants (many of them, not surprisingly, alpines) will only germinate if they have had cold treatment. The usual procedure is to sow in autumn and place them outside during the winter, or to give them six weeks or so in the fridge before bringing them to room temperature.

What next?

Once your little seedlings miraculously appear, the standard horticultural procedure is 'pricking out' – transplanting your babies to a new home and better spacing. My own experience is that perfectly good results can be obtained by leaving your seedlings where they are, but thinning them. This feels like infanticide, but (unless you are growing on a large or commercial scale) there is nearly always too much seed in a packet, and getting rid of some is no great loss. Massive losses occur in nature and, if you are sowing outside, thinning is the usual procedure. When 'pricking out', or transplanting to their garden positions, remember about delicate stems and leaves – try only to handle by the seed leaves or by digging out the whole tiny plant and lifting it on a teaspoon to its new home. Remember watering in (see Chapter 2, page 24).

Spacing

A great deal of research has been done into the best spacings for different seedlings. This, I suspect, is largely because it can be done by experiments easy to control. Fortunately the instructions are usually on the packet – so follow them!

Hardening of

Baby plants reared in controlled conditions need to be acclimatized to the rigours of life outside – so put your seed trays or pots out for a few hours during the day for a week or two. Hardy varieties can then be planted out in their final positions, but tender ones need to wait until the danger of frost is past.

Protecting delicate seedlings

It's all very well putting your precious seedlings outside to harden off, but what about slugs and snails? Disaster can easily strike. The best remedy I have come across is the moated-table method. Get hold of an old table (from a junk shop or the tip) and place the legs in tubs of concentrated salt solution. Your seed trays can then safely sit on the table, because even the most daring of molluscs would fail to negotiate the salty moats. Planted-out seedlings are not so easily protected (see Chapter 6 for possible solutions to slug problems).

Old wives' tales – true or false?

'Old wives' tales' to do with sowing abound. Many of them are the result of generations of sensible observation and trial and error. Others are more dubious. Here is a selection.

Sowing by the phases of the moon

Since the moon has such a strong influence on movements of the oceans, adherents to this school of gardening believe it must also have an influence on events in the soil, on the weather, or perhaps even on living tissues. This is a perfectly reasonable supposition – the snag is that the exact nature of these events, and how any hypothesis might be tested, is extremely complex. Folklore has it that one should sow only with a waxing moon or at full moon, and never during a waning moon. Is this because at spring tides water rises very high and might do so in the soil too, providing more moisture for germination? But at *low* spring tides – which happen on the same day as the highest tides – the water goes *down* very low! This would then *deprive* seeds of moisture, so the argument seems illogical. There is some evidence that rainfall tends to be higher just before and at a full moon. Is this the basis for the practice?

Some experiments which have been carried out, and literature exists on the topic (see Further Reading), but the work needs to be repeated and extended before 'moon gardening' can be said to have a sound scientific basis. If you read the book by Kollerstrom do so with a critical mind.

Zodiac signs

The signs of the zodiac are supposed to be meaningful for gardening. If you study such a guide it actually makes good sense. Cancer (22 June–22 July) is reputed to be the most productive sign for sowing and transplanting (perhaps a little late for sowing!). Leo and Virgo (23 July–23 September) are 'barren signs' and not good times for sowing (well, yes, it's often too hot and dry). Libra (October), a 'cardinal air' sign and Scorpio (November), a 'fixed water sign' are supposed to be good times for sowing flower seed. This makes sense, as it is still reasonably mild, with enough light and moisture for young seedlings to get going if germination occurs (only advisable for hardy species though!). Autumn is, in fact, the time when seeds of many wildflowers are shed and, if mild enough, they often germinate and overwinter. Saggitarius (December) and Aquarius (January) are unfavourable – understandably. But the following Capricorn – a 'cardinal earth' sign – is supposed to be a good time to plant and sow 'those plants which produce an abundance of roots and branches'. This is a bit of a puzzle; does it refer, perhaps, to those seeds which germinate after stratification, which is the case for many tree seeds? My verdict is that, over the centuries, gardeners have experienced the influence of the time of the year on various processes and have linked these with the zodiacal calendar in a perfectly common-sense way. But I don't think astrology is necessary for working out when to do things.

The 'bottom test'

It is reputed that the truly dedicated farmer used to drop his trousers and sit bare-bottomed on the soil. If it felt comfortable then it was time to sow – eminently good sense. I'm not recommending this – but an 'elbow test' when sowing out of doors could well be useful.

As the debate about global warming continues, farmers may find the soil uncomfortably hot on the bottom in a decade or two – but by then gardening practices will have altered dramatically, anyway.

Sow generously

One for the rook and one for the crow
One to die and one to grow.

Yes, indeed!

Soaking seed in manure tea

A method which is supposed to give rapid and successful germination is to prepare a bucket half full of manure and half water, let it stand for a day, strain and dilute it until it acquires a weak tea colour, and soak seeds in it for a few hours. This practice would lend itself to experimental testing! Take several species and set up trials using at least 30 (preferably more) treated seeds to compare with untreated ones. All other conditions must be kept constant (the tricky bit); the conditions of light, amount of watering, temperature and so on must be identical. Scientists usually talk about 'germination rates' and 'percentage viability' when recording the results of such experiments.

Creating scented flowers

Folklore has it that if you soak seeds of a species which hasn't much scent of its own in a sweet-smelling solution, the flowers produced will have the identical scent. How does this tie up with treating peas and beans with paraffin to ward off mice? Do the flowers then smell of paraffin? Try it and see!

Figure 3-3. Different kinds of meristems.
(a) Apical meristem, for example at the tip of a shoot.
(b) Meristem in the cambium, the layer of cells between xylem and phloem in a stem for example (in longitudinal section).
(c) Meristem at the node of a grass stem – where a leaf branches off.

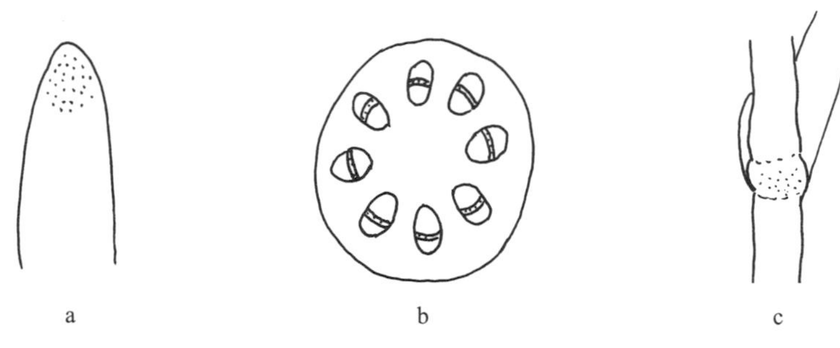

Growing up

There are similarities and differences between the way we, as humans, grow and the way plants grow. The main difference is that plants grow only at clearly defined 'growing points' called **meristems**. The similarities are that growth is influenced by nutrition and controlled by hormones.

Meristems occur at the very tips of leading shoots, in the tips of lateral buds and at the tips of roots. There are also meristems inside stems and roots and at intervals up stems (see above). In these growing points cells are incredibly active – dividing rapidly and busy making new cytoplasm and cell walls. They therefore need supplying with plenty of energy and the materials with which to make new tissues.

Food for growth

The requirements for good growth are explained in Chapter 1 (see page 12).

Your soil must contain the right chemicals for whatever you want your plants to do – provide flowers, vegetables, fruits or seeds. It is your choice whether to nourish your plants with artificial man-made fertilizers, whether to rely on material recycled from dead plants and animals, or to employ a combination of both. Chapter 4 deals with the soil and how to keep it in good shape to provide everything plants need for optimum growth and health, attractive flowers or productive fruit and vegetables.

Control of growth – plant hormones

How is it that shoots always grow upwards, bend towards the light, and sometimes twine round supports, while roots always grow downwards? What determines when and how buds come into leaf, your lettuces 'bolt', or autumn leaves change colour and fall? These events and many others are now known to be controlled by plant hormones, sometimes simply called **'growth substances'**. There is still a lot to be learnt about how they work, and the subject is immensely complicated. They are specialized chemical substances produced mainly in growing regions of the plant, but they can be moved around in solution to other sites where they have their effects. What follows is a simplified whistle-stop tour of the main substances, their functions and practical applications for gardening. The first three represent groups of several related chemicals. Abscisic acid and ethene, on the other hand are single, identifiable substances.

Auxins

One of their important functions is to promote cell enlargement and, in shoot and root tips, they bring about growth by increasing the length of cells just behind the region of cell division. They also initiate root development from calluses and cuttings and play a part in fruit setting and growth. Another function is to promote growth at the apex of a stem and prevent excessive growth of lateral shoots – **apical dominance** (see the tips on pruning, page 59).

Gibberellins

They promote cell division – production of new cells rather than enlargement of existing ones – but they also encourage increase in size of cells. In some plants they are responsible for bolting – the exceptionally rapid growth of flowering stems. They are also implicated in bud-breaking and the breaking of seed dormancy.

Cytokinins

They promote cell division in stems but, although produced in roots, are inactive there and actually have an inhibitory effect on the growth of roots. They act very much in conjunction with auxins, and the proportion of auxin to cytokinin is very relevant; a high cytokinin to auxin ratio favours shoot growth, whereas high auxin in proportion to cytokinin promotes root growth. Like gibberellins they break bud and seed dormancy, and they also delay leaf senescence – easily demonstrated by a simple experiment in which a drop of a cytokinin placed on a dying leaf results in a green island of living tissue in an otherwise yellowing blade.

Abscisic acid

Unlike the three described above this substance is an inhibitor of growth and, in this capacity, is important in protecting plants from stress. For example, it promotes the closing of stomata and inhibits stem growth during drought conditions: it promotes bud and seed dormancy; and is responsible for sealing off the connection between leaves and their stems prior to leaf fall (formation of an abscission layer – hence the name of the hormone).

Ethene (formerly called ethylene)

Unlike the previous four this is a gas which has the effect of promoting fruit ripening (hence placing a ripe banana skin amongst green tomatoes) and the senescence of leaves and flowers.

How do plant hormones work?

This brief summary begs many questions, amongst them how these substances are produced in cells, what determines when and how they are secreted, and how they get to the places where they have their effects. There is a lot still to be discovered, but some simple experiments have provided a few clues as to what goes on in response to outside stimuli such as light and gravity.

How do shoots always bend towards light?

The earliest work which led to the discovery of auxins were follow-ups to some simple experiments devised by Charles Darwin and his son Francis. Curious to find out how it was that shoots grew towards a source of light they tried removing the tips of the shoots of oat seedlings, and also covering them up. Figure 3-4 (overleaf) shows the results. From this they concluded that there was some 'influence' produced in the tip which had an effect on a growing region just behind.

In 1928 a Dutch scientist showed that it was a chemical substance which was responsible for the effect. He had the idea of using small pieces of jelly infused with 'juice' from the tip of a normal shoot. Figure 3-5 (overleaf) illustrates what he did.

The question of what exactly goes on when light strikes the cells on one side is still not fully answered. How do the cells 'know' that light is shining on one side only and that it is time to release auxin on the opposite side? The exciting thing about science is that solving one problem simply leads on to more questions!

How do roots always grow downwards, and shoots upwards?

Imaginative experiments have been devised using a piece of apparatus which rotates a germinating seed in a horizontal position so that the effects of gravity are equalized on all sides. Both the shoot and root grow horizontally. But stop the rotation and normal response to gravity is resumed. Further experiments with shoots have shown that the response is due to unequal distribution of auxin; in a horizontal shoot, auxin appears to move downwards and therefore stimulates

Figure 3-4. Diagrams to illustrate the principle behind the experiments carried out by Charles Darwin to test his hypothesis that something in the shoot tip enables it to bend towards the light. The arrows represent the direction of light.
(a) Normal shoot illuminated from above.
(b) Normal shoot illuminated from one side grows towards the light source.
(c) Shoot with tip cut off.
(d) Decapitated shoot covered with lightproof cap
(e) Shoot with main part of stem buried in sand.
(c) and (d) showed that when there is no tip there is very little growth and no bending towards the light, supporting the idea that the determining factor is in the tip. In reality this experiment would be done with several shoots getting each treatment and there would be control shoots (normal shoots without the treatments).
(Adapted from Soper [ed.] *Biological Science* Vol 2, Cambridge University Press, 1984, by permission of CUP)

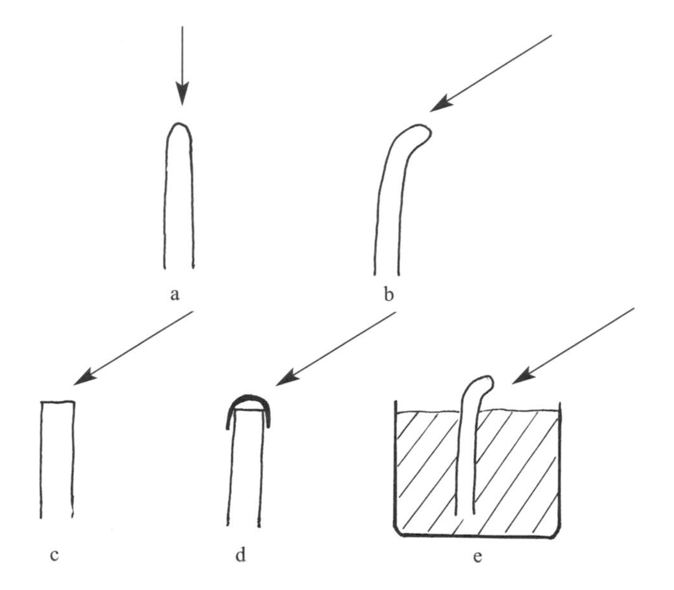

growth on the lower side so the shoot curls upwards. Something similar probably happens in the root, though experiments have been somewhat inconclusive. An intriguing thing about roots is that cells in the root cap contain large starch granules, and the hypothesis has been put forward that these sink to the lowest part of any cell under the influence of gravity and so give a signal which triggers an appropriate distribution of hormone. But exactly how this operates is still something of a fascinating mystery.

Other scientists went on to conduct similar experiments, followed by more sophisticated ones involving culturing plant tissue in the laboratory, to unravel the story in more detail. One classic test showed the interdependence of auxins and cytokinins. There is still a long way to go in fully understanding how plants respond to outside influences.

Figure 3-5. Diagrams to illustrate the principle behind further experiments with shoot tips.
(a) Shoot with tip removed but no treatment.
(b) Shoot with a piece of agar jelly which has been treated with an extract of juice from another shoot tip.
(c) Treatment with agar jelly without shoot-tip extract.
(d) Treatment on one side of the decapitated tip only.
These results show that a chemical substance appears to be speeding up growth on one side of the shoot to make it bend over.
(Adapted from Soper [ed.] *Biological Science* Vol 2, Cambridge University Press, 1984, by permission of CUP)

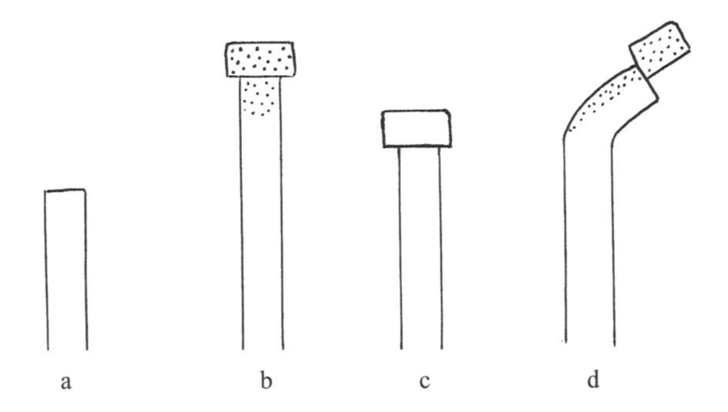

Etiolation

We're only too well aware that forgetting to bring germinated seedlings out into the light results in long, lanky pale stems. This is a natural response in which the plant grows upwards rapidly in an 'attempt' to find the light it needs. This is all to do with chemical reactions involving an important substance called **phytochrome**. (To be strictly accurate the word embraces a group of substances as there are several different kinds of phytochrome). Phytochrome is a pigment which absorbs light and, according to the intensity of light and the proportions of different wavelengths (particularly of red and infra red), it can 'tell' a plant whether or not to lengthen its stem or to make its leaves grow larger. It is also involved in the control of flowering and even of germination. Probably because of its importance in commercial horticulture this subject has received a lot of scientific attention (see Further Reading).

Practical applications

For domestic gardeners there are few practical applications to do with hormones - one obvious one is the use of rooting powders in taking cuttings. But for commercial horticulturalists there are several, and some of these are discussed below. In the future hormones may be useful as weedkillers, though more research is needed (see Chapter 5, page 97).

Propagation without sex but with hormones to help – vegetative propagation

There are plenty of books giving advice on the practicalities of vegetative propagation – taking cuttings, layering, grafting and the treatment of bulbs, corms and rhizomes and so on. Here I shall only describe some of the specialized organs which plants have to help them reproduce asexually (see Figure 3-6). I'll also touch on some of the fascinating and sophisticated things which horticulturalists can do as a result of understanding how plant hormones work.

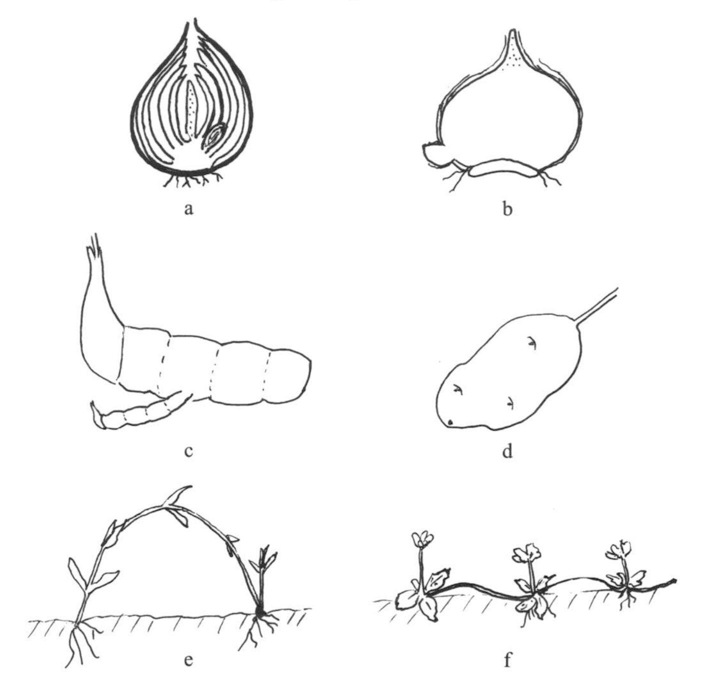

Figure 3-6. Various natural methods of vegetative propagation.
(a) Bulb, formed from the swollen bases of leaves with a new bulb forming (for example, daffodil).
(b) Corm, formed from the swollen base of a stem to form a globular structure, the shrivelled remains of the previous year's corm being visible at the bottom (for example, crocus).
(c) Rhizome, a swollen underground stem which grows horizontally, the dotted lines representing the limit of each year's growth (for example, iris).
(d) Tuber, the swollen tip of a stem growing underground, the 'eyes' representing lateral buds (for example, potato).
(e) Stolon, the tip of a stem which has bent over and rooted (for example, blackberry).
(f) Runners, stems growing out horizontally along the ground and rooting at intervals (for example, strawberry).
(a–c adapted from Soper [ed.] *Biological Science* Vol 2, Cambridge University Press, 1984, by permission of CUP)

Practical applications and gardening suggestions

Rooting powders for cuttings

High concentrations of auxin at the bare surface of a cutting promote the formation of roots, but lower concentrations are required for the subsequent growth of these roots. It is important, therefore, that whatever is applied can degrade naturally. Synthetic auxins are effective in initiating root formation but they don't degrade in the soil, and it has been found that the best substance is a naturally occurring auxin, indolyl butyric acid. This chemical is now the ingredient in most rooting powders. But it isn't a simple matter: not all plants behave as you would wish, and rooting is also influenced by other factors such as day length and the age of the plant tissue used. As with many aspects of the internal workings of plants there is still a lot to learn. An encyclopaedia of gardening will tell you whether a plant will root easily from a cutting.

Juvenility

It's not commonly realized that plant tissues go through a definite transition from childhood to adulthood. After a plant has reached a certain size a change occurs at the growing points and physically different adult tissue is produced. Experiments with ivy (which shows marked structural differences between its young form and its grown-up state) have shown that hormones are involved – adult plants can be made to revert to a juvenile form by treating them with gibberellins. Whilst this is not common practice in horticulture, cutting back of new growth is used, for example to keep woody plants young (the bases of plants are invariably still in the juvenile phase) and to provide suitable material for hardwood cuttings.

Changing plant shape

Weird and wonderful things can be done to produce specialist plants. Gibberellins are used to produce elongated stems, for example in growing standard fuchsias. Conversely, the popular ornamental pot plants, such as poinsettias and chrysanthemums, are kept small by treatment with growth retardants. These are synthetic substances which suppress the production of gibberellin. The chemical treatment doesn't last indefinitely, so if you plant a commercially grown poinsettia outside it is unlikely to remain in its well-disciplined form.

Micro-propagation

Some of the multitudinous, identical-looking and healthy plants on sale in garden centres start life as a tiny fragment of tissue cut from a parent plant and reared in a laboratory dish. They are the cloned 'in vitro' babies of the plant world. (See Chapter 7, page 140, for cloning and micro-propagation.)

Storing cut flowers and vegetables

Ethene is a gas produced by ripening fruit. It is therefore useful in encouraging ripening, but it is also produced by ageing tissues and it is therefore detrimental to cut flowers intended for sale as fresh as possible. Research has thrown up certain chemicals which inhibit the effect of ethene and these are used to treat flowers before packing. Another tip is to avoid storing apples, pears and bananas (ethene producers) in close proximity to green vegetables, particularly at room temperature.

Control of flowering

Hormones are almost certainly involved in the promotion and timing of flowering, but the situation is very complex and by no means completely understood, influenced by other factors such as day length, and variation from species to species. There is, curiously, one group of plants, the Bromeliads, which can be treated by ethene to induce flowering. This is yet another of the mysteries which still awaits a scientific explanation.

Slowing down and senescence

Under stress and unfavourable conditions plants must slow down their growth or even become dormant, and annuals have only a short lifespan once they have reproduced. Two plant hormones are very important in slowing growth and initiating dormancy and senescence: **ethene** and **abscisic acid** (the latter commonly known as ABA).

As far as stress is concerned there is evidence that abscisic acid promotes the closing of stomata in wilting leaves and therefore slows down water loss, and also photosynthesis by restricting the supply of carbon dioxide.

As winter approaches in temperate latitudes, or a dry season in the tropics, a layer of cells at the base of each leaf begins to break down (the **abscission layer**). Immediately below it a protective layer is formed which is waterproof and prevents entry of infective agents. The same happens at the base of the stem of a ripe fruit. Eventually the leaf or fruit falls off and, in fruits, there is evidence that the process is controlled by abscisic acid. This has commercial significance because ABA can be sprayed onto fruit crops to synchronize fruit drop. For leaves it appears that ethene is a controlling factor, and it is certainly involved in the ripening and gradual senescence of fruits. So, again, this means that commercial fruit producers can manipulate the speeding up or slowing down of fruit ripening either by applying ethene or a substance which inhibits ethene's action.

As already emphasized, the control of growth processes and the interactions of hormones are extremely complex, and there are few things an ordinary gardener can do to manipulate them apart from providing the optimum conditions for healthy growth. But there is a tip for tomato growers, and it is worth talking a little about pruning, since cutting bits off does affect growth patterns by changing hormonal balances.

Gardening suggestions

Tomato ripening

It is common knowledge that green tomatoes can be ripened by keeping them enclosed in a box or drawer. It helps, however, if a fully ripe one is with them from the start, because it will emit ethene and stimulate ripening in the others. Even better is a ripe banana skin. Some tomato growers even drape banana skins over their plants to encourage ripening!

Pruning

A characteristic of plant growth is that the topmost shoot (leading shoot) exhibits dominance over the lower shoots, which grow more slowly and laterally, rather than upwards. The phenomenon is called apical dominance and was mentioned earlier in connection with auxins which are thought to be involved in maintaining the effect. In other words the leading shoot pushes on upwards and the lower shoots are discouraged from growing too much. If, however, the growing point of the leading shoot is removed, apical dominance disappears and lower branches get a chance to grow more vigorously. Thus to obtain a rounded, bushy shape the leading shoot should be pruned back, plus any subsequent leaders which may take its place. (See also tree pruning in relation to protection of sheltered branches, Chapter 2, page 34.)

Another feature of growth is that a plant will tend towards maintaining a constant ratio of shoots to roots. Thus hard pruning does nothing to reduce size, rather it encourages more vigorous growth. For a balanced shape the rule is therefore to prune vigorous shoots lightly and less vigorous ones harder. If you want to avoid too dense a structure, remove some shoots completely rather than tinkering with chopping small bits off. The constant shoot-to-root ratio explains the practice of root pruning to create bonsai trees - less shoot must be balanced by less root if the tree is to be kept small. Unfortunately root pruning is not really practical if you want to keep a garden shrub or tree from growing too tall - hence the constant battle to hold back the jungle!

There is no shortage of detailed advice on pruning in gardening books.

Common growth problems

How frustrating it is when our beloved plants grow too slowly or too fast, produce too much leaf and not enough flower, or even flower furiously but then give up the ghost. Here are a few tips.

Gardening suggestions

Growth too rapid

This is invariably due to lack of light. Lanky seedlings need to be given a higher light intensity, preferably from above so that it is evenly distributed.

Growth too slow

This could be due to low temperature or lack of light, slowing down photosynthesis or, ironically, too high a temperature which acts to speed up respiration (loss of sugar) more than it is speeding up photosynthesis (gain of sugar). It could also be due to lack of nutrients and/or water. Try to analyse the situation in relation to all these factors.

Too much leaf and poor flowering

Invariably due to too much nitrogen and insufficient potassium. Don't overfeed with manure or general fertilizer.

Furious flowering (of shrub) and subsequent deterioration of plant

There could be some kind of stress – a disease, or prolonged drought or nutrient deficiency. Excessive production of flowers is an adaptation to ensure maximum reproduction and passing on of genes before the plant dies. In other words it is a 'cry for help', so try to find the stress factor and alleviate it.

4. Looking after the soil – the balanced approach to its care and maintenance

Is soil necessary?

You'll remember the discussion about the importance of roots being bathed in a watery solution. So why should gardeners bother with soil at all? It is, indeed, possible to grow plants in solution, a procedure known as **hydroponics**. Maybe one day genetically-engineered people will grow genetically-modified plants in artistically-designed trays containing scientifically-predetermined solutions. But it's pretty clear that the expenditure of energy required would, in present circumstances, be prohibitive!

Soil is arguably our planet's most precious resource, and it is not as abundant as we might think. This was brought home to me recently as I was gazing at England's famous white cliffs of Dover from the sea. There is a pathetically thin brown line above the white chalk. When you're next near cliffs take a look and I guess you'll see something similar. Alas, we're abusing that thin layer of soil dreadfully. An alarmingly large amount of soil gets eroded away into water courses, or blown away in dust storms, bad farming practices being invariably responsible. The pity of this despoliation is captured poignantly in the words of the great American conservationist, Aldo Leopold, speaking of a farm in Illinois:

> *Everything on this farm spells money in the bank... In the creek bottom pasture, flood trash is lodged high in the bushes. The creek banks are raw; chunks of Illinois have sloughed off and moved seaward. Patches of giant ragweed mark where freshets have thrown down the silt they could not carry. Just who is solvent? For how long?*

(Leopold, Aldo. *A Sand County Almanac*, Oxford University Press, 1949, reprinted with permission of OUP)

So, as gardeners working with soil, we have a special responsibility to care for it. We have to cope with its many quirks and uncertainties, and I hope this chapter will help to reduce some of the difficulties and frustrations.

It's vital to emphasize that soil is a whole ecosystem in its own right. Ecologists use the term 'ecosystem' to mean the interrelating network of living and non-living features of the environment. We tend to focus our attention on the above-ground parts of our gardens, but look below the surface and there is an awful lot going on. In a healthy soil there are myriads of tiny creatures beavering away, eating up all the dead bits and pieces which fall to the ground or break off from root systems. It's thanks to them that, on country walks, we aren't up to our necks in years of accumulated autumn leaves, animal faeces and dead bodies. They are the waste managers of our planet *par excellence*, and make it possible for luxuriant vegetation to keep going, despite receiving no input of nutrients from outside. Tropical forests, for example, could not exist without the soil organisms which recycle materials for re-use by the trees (see Woodland, Chapter 6, page 109).

Scientific research

A vast amount of work has been done – soil samples are easy to extract, compare, examine under the microscope, do controlled experiments with and so on. It is also possible to set up plots of soil out of doors and experiment with them, though it is more difficult to make conditions exactly the same, while varying one or two factors, than in a laboratory. In the UK a lot of this research has been done at the Rothamstead Institute where plots of soil have been studied over many years. What follows is a summary of some of the knowledge scientists have gained.

Soil content

- **Mineral particles** – of different kinds and sizes.
- **Water and dissolved mineral salts** – in films around the particles.
- **Air** – in spaces in between the particles.
- **Dead organic matter** – which, when well broken down, is called humus.
- **Living organisms** – from large ones such as moles, through medium-sized earthworms, insects, molluscs to microscopic fungi and bacteria.

Since the characteristics of different kinds of soil are largely determined by the properties of the mineral particles it makes sense to begin there.

Soil types

Chapter 2 touched on the differences between sandy and clayey soils (see Figure 2-2, page 21). Table 4-1 and Figure 4-1 show a little more detail.

Table 4-1. Characteristics of different types of soil.

Particle diameter	*Pore size*	*Characteristics*	*Type*
0.05–2 mm	Large	When dampened will not stick together in the hand. Well aerated but heats up and dries out rapidly. Poor in nutrients.	Sand
0.002–0.05 mm	Medium	When dampened will stick together to form a 'sausage'.	Silt
Less than 0.002 mm and flattened	Very tiny or non-existent	When damp is glutinous and can be rolled into a sausage and then into a continuous clay ring. Drains poorly and becomes waterlogged and cold in winter, solid with cracks when dry.	Clay

A simple experiment shows what proportions of the different kinds of particles you have in your soil (see Figure 4-2 overleaf). Find a narrow, transparent jar or bottle, the narrower the better. Put soil up to about 8 cm into the jar (more if it's a tall one), add water to about 3 cm above the level of the soil and shake vigorously. Leave the soil particles to settle completely, which may take two or three days. You'll see the different-sized particles in layers: the largest, heaviest, sandy ones at the bottom where they have settled out first, then medium-sized silty ones. Finally the clay particles appear on top and, if the water is still cloudy, it may be because there are very tiny ones remaining in suspension. You can get a rough idea of your type of soil from the relative widths of the layers.

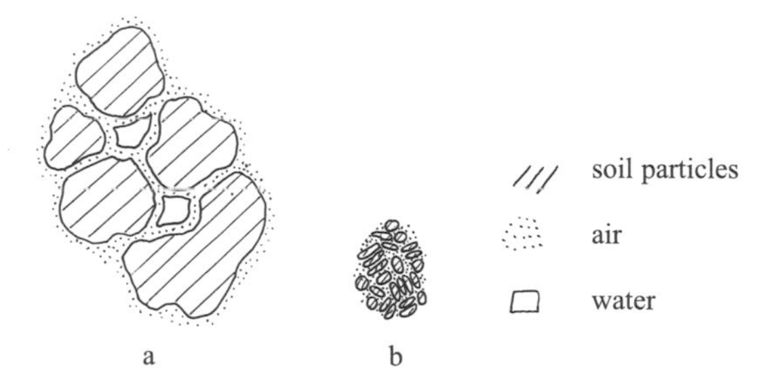

Figure 4-1. Water films in different soils (not to scale). Because of the surface tension effect water clings to particles, so that each grain is surrounded by a thin film of water. (a) sandy soil: large, irregularly shaped particles don't fit together closely, so the water films are not continuous and there are air spaces through which oxygen can diffuse rapidly and through which water can flow; (b) where the particles are very tiny, as in clay, they pack so closely together that the water films are continuous and there may be no air spaces – hence waterlogging and shortage of oxygen. This is a simplification of water films – scientists love distinguishing different grades of particle size, different measurable tensions and so on, but as gardeners this is really all you need to know.

Figure 4-2. Possible results of an experiment in which soil is shaken up with water.

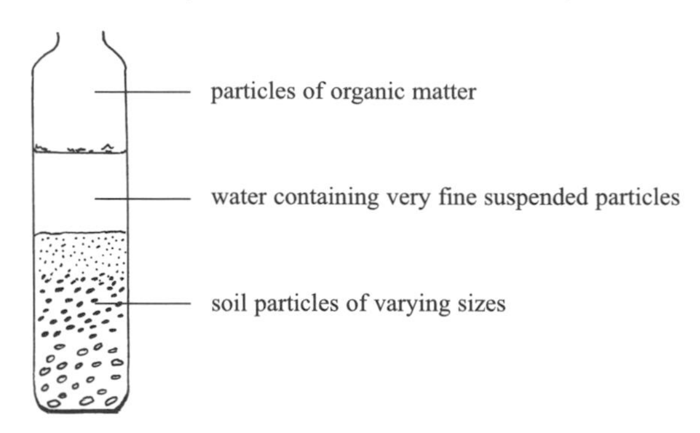

The best kind of soil for a garden is loam, a mixture of sand, silt and clay particles in approximately equal proportions. In addition there should be some organic material. This produces a soil which is reasonably well aerated and drained, but with adequate water retention. There is, of course, a spectrum of different soils from pure sand at one end to pure clay at the other, with many intermediate types in between. Loam is the ideal to suit the widest range of plants but don't despair if you have one of the extremes – read on!

Providing the best conditions

In natural ecosystems, mineral nutrients reach the soil from weathering of underlying rocks, from materials brought from elsewhere – eroded by glaciation, wind or water – or by the breakdown of dead organisms by invertebrate animals (for example, molluscs, earthworms, millipedes), bacteria and fungi. Only those plants which are adapted to survive in the pertaining conditions will grow and flourish. But in a garden we have control over the situation and have many options; we can change the soil structure, regulate the water content and acidity or alkalinity (pH), or provide whatever nutrients we desire – either by emulating natural recycling of organic matter or by adding man-made fertilizers. The numerous books covering this tend to fall into two camps – those which are about so-called 'organic' methods, and those which are unashamedly in favour of artificial aids. Scientific evidence points, however, to a common-sense approach which is to make the best of both. My own philosophy is to do just this. What follows is an outline of what might be dubbed 'integrated soil management'.

Managing soil structure and texture

If you are blessed with a loam soil be thankful. If your soil is at the clayey or sandy end of the spectrum, and you wish to grow a wide range of plants, you may have problems which require some pretty hard work to solve. Below are some management suggestions.

At the extremes of clay or sand you may have to compromise. Rather than trying to alter your soil a simpler solution is to find plants which can cope with the extreme conditions.

Gardening suggestions

Very sandy It is difficult to alter the particle content, but the addition of plenty of organic matter will help to make the soil less prone to drought. Without large amounts of humus (well-rotted organic material) the coarse particle structure lets rain water drain through very quickly, taking with it dissolved minerals. This '**leaching**' means that the soil needs regular replenishment with nutrients. Add plenty of compost or well-rotted farmyard manure every year to help to reduce leaching and add nutrient elements.

Very clayey Add horticultural sand or grit (not builders' sand) to help separate out clumps of clay particles and improve drainage and aeration. A lawn, for example, can first be spiked and sand scattered into the holes. Dig beds over with generous helpings of sand or grit. In vegetable and flower beds the addition of organic material helps, too.

To dig or not to dig

There are diametrically opposed schools of thought on digging policy, and it is difficult to find good scientific evidence to support the different ideas. Physical stirring up must influence the structure of the soil by distributing materials, improving drainage and aeration, and killing weeds. Digging and raking also creates suitable seed beds, making life easy for young roots which have to be able to push easily between the soil particles. So it makes sense to do some light digging with a fork in spring. Solid clods left on the surface should break down naturally by frost action. Double digging (going down to two spades' depth) is traditionally meant to improve the availability of nutrients, and encourage deeper root systems, but it probably expends more effort than it is worth - though I have no scientific evidence to support this statement.

Compacted soil is going to be less well aerated, so avoid walking on the soil as much as possible. Try to design vegetable beds so that all parts can be reached from a path. Where walking on the soil is absolutely necessary, planks laid down help to distribute the load and lessen the intensity of pressure on small areas.

On the other hand, earthworms do a fantastic job in aerating and mixing the soil. The 'no-dig' school of thought relies on the natural activities of these and other living organisms to keep the structure of the soil in good heart. In theory this is a valid approach. In a woodland seedlings manage to become established without any human intervention in the soil. It is on this principle that permaculture is based – a system in which growing is usually done in a woodland kind of setting. As with so many gardening practices, it would be extremely difficult to design a fair test to compare the efficacy of digging and growing in bare ground with any other kind of approach. The problem, again, lies in standardizing all the variable factors. In the useful *Basic Book of Vegetable Growing* by W.E. Shewell-Cooper (1972) the author claims to have made a scientific comparison between 'dig' and 'no dig' over a period of 10 years. His conclusions were that 'no dig', covering the soil with a thick layer of compost and leaving well alone until a light forking in the spring, produced better results. He doesn't explain his methods, though – something one would like to know before deciding whether the research was valid. Proponents of permaculture claim good results but, from my reading, I gain the impression that it only lends itself to growing certain types of plants – a pretty flower bed with both perennials and annuals wouldn't be easy to achieve. Perhaps one advantage of not overdoing the digging might be that there will be less disturbance to the beneficial fungi which are associated with many plants' roots (see Chapter 2, page 22).

But *do* encourage earthworms! If you reckon your garden is short of worms they can be purchased from 'worm farms'. But you must supply them with plenty of organic matter to feed on, make sure their home doesn't become too acidic, and keep them warm in winter.

What to do in a drought

Again, policies are a matter of opinion and these seem to differ, with little good scientific evidence to support them. To hoe or not to hoe? I have seen it suggested that disturbing the surface is a bad idea because it allows more evaporation of water from below. But I have also come across advice to hoe the surface during dry weather in order to create a 'dust mulch', which prevents water loss. You can carry out your own experiment because the situation is fairly simple. Find two bare plots with, as near as possible, identical soil. Look out for a weather forecast of several dry days, then hoe the surface of one plot so that it dries out to form a good thick dust layer on the surface. Measure the humidity in the two plots at the same depth, below the dust layer, and at regular intervals. Instruments for this purpose can be obtained from garden centres (and are also useful in the house to check whether potted plants need watering). Your measurements should tell you whether or not there is any marked difference between the two treatments.

Application of a mulch of organic material certainly helps to retain soil moisture, but it has to be thick (say 10 cm) and evenly spread. It must be applied when the soil is still damp. Inevitably most such mulches will, before long, be incorporated into the soil by worms and bacteria. A good one is chopped bark, which is not so rapidly broken down.

Another way of ensuring minimum evaporation from the surface is to plant thickly so that there is very little bare soil exposed to sun and wind. You put your washing out on a sunny windy day for *maximum* evaporation! You want to achieve the opposite for your soil surface.

'Feeding' - fertilizing

Ideas about providing soil with the correct nutrients vary. I shall look at the two extremes – using artificial, man-made fertilizers *vis-à-vis* 'organic' methods – separately. Then I shall make a case for combining the two.

Fertilizing using man-made chemicals

Table 4-2 shows what scientists have discovered so far about the functions of the main elements required by plants (those in addition to carbon, hydrogen and oxygen, obtained via photosynthesis – see Chapter 1). When gardeners talk about 'feeding' these are the items on the menu.

The information in Table 4-2 has been obtained mainly by the kind of experiment in which batches of plants are *deprived* of one particular element and a record kept of what happens to them. For example, plants deprived of magnesium go yellow and this information, combined with a knowledge of the structure of chlorophyll obtained by standard chemical analysis, provides good evidence that magnesium is essential for plants to make chlorophyll. Similarly we know that nitrogen is one of the elements of which proteins and nucleic acids are made, and phosphorus, too, is an important component of DNA and RNA. This ties up nicely with the observation that plants just don't grow properly without them.

The functions of some elements, though, are not so thoroughly understood, for example, potassium. In animals – including ourselves – potassium is known to be involved in electrical changes in membranes which facilitate the transmission of nerve impulses. It is also known to be a temporary co-enzyme and facilitator of various reactions. Experiments on plants have shown that, when deprived of potassium, strange things happen to the leaves. To the best of my knowledge, though, the effects on flowering have not been observed and recorded in the scientific literature. All gardening books, however, describe potassium

Table 4-2. The main elements and what is known of their functions (see also Chapter 1, Table 1-1).

Macro-nutrients (main elements)	
Nitrogen	Component of proteins and other important compounds such as chlorophyll. Deficiency results in stunted growth and yellowing of leaves. Important for general growth, particularly of leaves.
Phosphorus	Component of nucleic acids and some proteins. Also important in phospholipids which are essential for the structure of cell membranes. Deficiency results in stunted growth, particularly of roots. Important for general growth and good root formation.
Potassium	Known to be involved in the active transport of substances through cell membranes and is thought to act as temporary activator for many enzymes. Deficiency results in browning and curling of leaf margins, often on lower leaves first. Has always been reputed to encourage flowering, though the scientific understanding of this is uncertain.
Sulphur	Synthesis of proteins and many other organic compounds. Deficiency results in leaf yellowing and sometimes deformity. Important for general health and growth.
Magnesium	A component of chlorophyll. Deficiency results in yellowing. Essential for photosynthesis and therefore for healthy growth
Calcium	Important component of the 'cement' between cell walls. Deficiency results in stunted growth and blackening of young leaves. Essential for the construction of cells and therefore for normal growth.
Micro-nutrients (trace elements)	
Iron	A component of certain enzymes involved in respiration and photosynthesis. Deficiency results in yellowing of older leaves and almost complete loss of colour in young leaves.
Manganese	A component of enzymes involved in respiration and photosynthesis. Deficiency results in discoloration of new leaves. Needed in small amounts for general healthy growth.
Copper	A component of certain enzymes. Deficiency results in die-back of shoots and discoloration of leaves. Needed in small quantities for continuing healthy growth. Needs vary with species.
Molybdenum	A component of an enzyme involved in amino-acid synthesis. Deficiency results in slightly retarded growth and, in cauliflowers, a strange twisting of the leaves. Needed in small quantities for healthy growth. Needs vary with species.

Micro-nutrients (trace elements)	
Zinc	Is known to be involved in anaerobic respiration. Deficiency in some plants, for example, *Citrus,* leads to leaf mottling. Needs appear to be specific to species.
Boron	Deficiency results in abnormal growth and death of shoot tips and in some species cracking of tissues. Appears to be involved in cell division in meristems. Needed in small quantities for normal growth and needs vary with species.
Chlorine	Deficiency can result in wilting. Its role is not well understood.

as being important for good flowering and I can only surmise that this is based on generations of practical experience. Since flowering is known to be influenced by plant hormones which have to be passed from cell to cell across membranes it may be that potassium is involved in this process. That is my hypothesis, but I am not necessarily sticking to it until I can think of a way of testing it – a very technical matter involving sophisticated laboratory procedures and as fraught with difficulties as investigating different cultivation regimes!

The table shows that nitrogen (chemical formula N) promotes good growth, particularly of luxuriant leaves; phosphorus (P) encourages healthy growth, particularly of roots, and potassium (K) good flowering. Hence the term for a general fertilizer is NPK. If you look at the constituents listed on a packet of a *general* fertilizer you will find the three elements in *very roughly* the proportions N 2:P 1:K 2 (7% N 3% P 6% K). These figures represent the actual elements. Most packets now give a breakdown of both the compounds in which the elements are present (for example, phosphorus pentoxide and potassium oxide), and the elements themselves, so look for the latter. You'll notice, however, that a fertilizer designed specifically for flowering, such as those advertised for tomatoes, has a higher proportion of potassium: N 2:P 1:K 3.

How does a forest flourish without any input of nutrients?

Think of a tropical forest – you may not have been into one but I guess you will have had a taste from TV programmes. The elements just described are essential for the prolific growth of those gigantic trees. But how is it that the nutrients in the soil are not depleted?

The answer lies in rapid recycling. Leaves and wood die and fall; they are consumed by creatures on the forest floor and the faeces of these animals are further broken down by even smaller creatures – in other words the plant material is decomposed. The forest floor is like a gigantic compost heap (see page 73 on compost making) and the high temperature and humidity make the process so quick that nutrients are returned to the soil as rapidly as they are absorbed by the trees. Tropical forests are amongst the most productive ecosystems in the world – but no one needs to go and fertilize them.

Do garden soils really need 'feeding' with artificial fertilizer?

In the world of intensively managed ecosystems things are different from virgin forest. In agriculture, and commercial horticulture, crops are grown on a large scale and regularly removed from the site. This can lead to gradual depletion of soil nutrients, and scientific studies have shown that the application of artificial fertilizers improves productivity. In an ordinary garden, though, such large losses from the system don't normally occur, so it's sensible to ask whether adding fertilizer is necessary. The answer is that situations will vary. Vegetable plots with heavy annual cropping are likely to benefit from fertilizing. Likewise pots and containers, from which soluble nutrients readily leach out, will need feeding. But a herbaceous border planted with perennials shouldn't need a great deal of attention as far as nutrients are concerned. Scientific analyses of garden soil samples sent a few years ago to a laboratory by members of the Royal Horticultural Society showed that almost three quarters of them contained sufficient phosphorus to supply the needs of vegetable growing. Of course, a lot depends on the past history of a site. Artificial fertilizers became very popular in the fifties and sixties, and the present state of our gardens may reflect the excessive use of man-made chemicals during those years. The moral is to proceed with caution. Over-doing things is a waste of money and resources and can also result in toxic conditions (see below).

Gardening suggestions

What to use when

Laborious research and development work by numerous companies has resulted in a plethora of choice. The shelves of garden centres are stocked with all manner of packets and bottles. My policy is simple: as well as using lots of compost or manure (important because of soil structure – see page 64), I add artificial fertilizer (in moderation) to plants that are heavily cropped or pruned, or grown in pots; potassium for those which need to flower, nitrogen and phosphorus for those which need good foliage or roots only.

If I'm particularly keen to encourage flowering (for decoration, or for vegetables that are derived from flowers) I use a high-potassium fertilizer described as being for tomatoes. For feeding flowery containers I also use a tomato fertilizer. For plants such as herbaceous perennials less fertilizing will be required. Useful also are slow-release pellets. However, I have freely draining soil and my main herbaceous border is bounded by a neighbour's leylandii hedge which must consume nutrients. When introducing new plants, therefore, I dig in a little rose fertilizer and once a year, in spring, I give my flower beds a very light dressing of pure potash (usually potassium sulphate).

If I'm interested only in leaf growth (grass, non-flowering shrubs, leafy vegetables) or root crops I use a fertilizer with higher proportions of nitrogen or phosphorus. Nitrogen-rich fertilizers include ammonium sulphate and ammonium nitrate; phosphorus ones are single or triple superphosphate and diammonium phosphate.

If you were to accuse me of not being *truly* scientific about this you would be right. I haven't any evidence, by way of a controlled experiment, to tell me whether I would have done less well without taking this trouble. Nor can I be certain that I wouldn't have done equally well by practising 'organic' methods. I derive a lot of pleasure from the results though, and visitors seem to like my garden. I base what I do on my scientific knowledge of the elements which plants need but, as I pointed out in Chapter 1, the life sciences are a bit fuzzy round the edges.

A word about nitrogen

This has a complicated turnover in the soil. The relationships between organic compounds in plants, mineral nitrogen (ammonium, nitrate and nitrite), nitrogen in the air and bacteria in the soil are tortuous, and making sure that nitrogen is available to plants is more complicated than managing other nutrients. This will be discussed in the section on organic fertilizers (see page 72).

Iron

It's also worth bearing in mind that some nutrients may be present in large enough quantities but, for some reason, are not in a form your plants can absorb. Iron is the classic example, becoming less easily absorbed by plants as the soil becomes more alkaline. Plants adapted to acid soils – such as rhododendrons, azaleas and camelias – are particularly prone to iron deficiency if grown in neutral to alkaline conditions. This can be corrected by applying a solution specially formulated to contain absorbable iron called chelated or sequestered iron. It is sold as a proprietary product along with other liquid fertilizers.

Trace elements – micro-nutrients

The elements under this heading are only required in very small quantities and tests have shown that, in most garden soils in the UK, there are adequate amounts of them. So there is no need to be concerned about supplying them.

A word of caution

Overdoing fertilizer application can be as deleterious as depriving your plants of nutrients. Very high concentrations of mineral elements can be toxic and can inhibit absorption of water by the roots (see osmosis, page 29). Too much nitrogen is also often the cause of shrubs producing luxuriant leaves but failing to flower. It isn't easy to get it exactly right – a lot depends on your type of soil, weather patterns, kinds of plants; for example, freely-draining soils tend to lose their nutrients more readily than water-retentive soils, some plants are more demanding than others (described by veteran gardeners as 'gross feeders'!). Unless you can afford to send soil samples to be analysed regularly the golden rule is to read the instructions on packets and abide by them.

Slow-release fertilizers

There is now a range of fertilizers in solid form which are designed to release their contents gradually – more like the way in which nutrients are released slowly during decomposition of organic material. They are particularly useful for containers and they also help to prevent over-fertilizing.

Foliar feeding

Fertilizer can be sprayed directly onto leaves as long as they are not hot and in direct sun. The solution is absorbed directly into the leaves. Care needs to be taken with dilution, according to instructions. To encourage climbers it is possible to spray walls with liquid fertilizer. The plants will happily absorb them and possibly grow more quickly.

Using organic fertilizers

The usual meaning of an organic fertilizer is one which is derived, or has been extracted from, material which was once living. First, some definitions:

- **Manure** – usually taken nowadays to mean animal droppings or urine mixed with straw, wood shavings or newspaper. Chicken manure usually comes as pure droppings. *High in nitrogen but contains little else.*
- **Compost** – the product of decomposed plant material, which may be garden waste or kitchen vegetable waste. (The word compost is also used to mean a growing medium.) *Usually high in nitrogen but other mineral content will depend on the plant material used to make the compost.* In theory you could sustain a garden by simply recycling everything into compost. In practice

that is not always possible (see 'Is it possible to rely entirely on recycling?', page 79).

- **Wood ash** – self explanatory. *High potassium content, low nitrogen and phosphorus.*
- **Bone meal** – ground-up bones from the livestock industry. *High nitrogen and phosphorus.*
- **Fish meal** – waste from the fishing industry. *High nitrogen and phosphorus.*
- **Others** such as dried blood, hoof and horn, seaweed and so on *have mostly high nitrogen, though seaweed can be a useful source of potassium.*

What happens in decomposing material, and why does a compost heap get hot?

Fertilizing with organic materials is a slower process than spreading or watering in an artificial fertilizer because it relies on the activity of living organisms to make the mineral elements available. One of the first things that happens in a compost heap is that it is invaded by small creatures, **detrivores** (sometimes, more sensibly I think, called detritivores) – slugs, millipedes, woodlice, earthworms – which like to feed on dead bits and pieces. They do a good chopping-up job and reduce material to even smaller particles which are eliminated in their faeces. Fungi and bacteria, meanwhile, have also colonized the plant material and they get to work on these smaller bits, too, and 'digest' them further. These micro-organisms have the ability to secrete enzymes into their surroundings (just as our digestive systems secrete enzymes into our stomachs and small intestines). The enzymes help to break down large organic molecules into simpler, smaller, soluble ones which the fungi and bacteria can absorb easily to feed themselves.

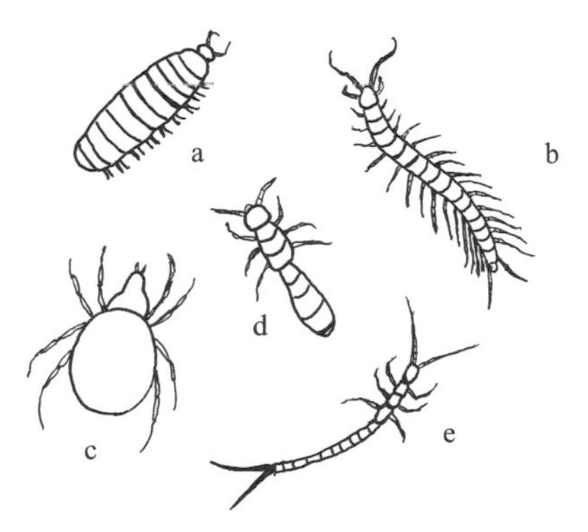

Figure 4-3. Some of the animals to be found in a compost heap or in soil or piles of dead leaves (not to scale): (a) pill millipede; (b) centipede; (c) mite; (d) springtail; (e) bristle tail. Springtails are very tiny but are immediately recognizable because, true to their title, they jump! There will also be woodlice, slugs, worms and other more familiar creatures.

They then excrete surplus or unwanted substances (mineral elements in soluble salt form) and these build up in the compost – ready to be absorbed by plants when the compost is applied to the soil. The relationships between all these busy creatures can be thought of as a network or web. Ecologists term it a **decomposer food web** (see Chapter 5).

If you haven't yet made the acquaintance of the inhabitants of your compost heap I can guarantee that you'll be intrigued. Equip yourself with a piece of white cloth – perhaps part of an old sheet – a paint brush, magnifying glass and trowel. Lay the sheet on the ground, shovel out a trowel-full of compost from well inside the heap and spread it out on the sheet. You should immediately see lots of small creatures scurrying around. But you have to be quick because, being soil animals, they don't like bright light and will dive for cover. The paint brush is to gently tease them out from wherever they are hiding!

A good compost heap is therefore a hive of activity – seething with living organisms from visible 'minibeasts' to invisible micro-organisms. But frantic activity needs energy. Respiration is the source of this (see Chapter 1), and respiration generates heat. You only have to think of yourself hot and sweaty after a spell of vigorous exercise. In only a day or two a well-constructed heap of garden waste may be literally steaming! This heat has a useful spin-off for gardeners as it acts to kill weed seeds. However, if allowed to get too hot, the process defeats itself because it kills some of the useful bacteria and fungi too and the rate of decomposition declines. At around 45°C weed seeds and disease-causing organisms are killed off, but above 55°C the 'goodies' begin to suffer. Large commercial compost heaps are usually kept at a temperature of about 60°C. In a garden, with regular addition of cool and varied material, excessively high temperatures are unlikely to be a problem. But your heap should be warm!

Composting is a splendid idea and a must if you have room, but don't expect things to happen quickly. It's a slow process, and one that doesn't always release large amounts of nutrients. Scientific studies on agricultural systems have shown that dependence on organic fertilizers results in lower productivity than is the case when artificial fertilizers are used. There are good reasons, though, for using manures and compost, to do with soil structure (see Gardening suggestions, pages 77-78) and the encouraging of biodiversity in gardens and the countryside in general (see Chapter 6).

Organic material and soil structure

The addition of organic material to the soil is so important that no garden should be without a compost heap or bin. It's also useful to find a friendly livestock farmer, or someone who keeps horses, to provide farmyard manure – not so easy these days when mixed farming is less common and so many large cattle farms do not have supplies of straw with which to mix slurry. Organic material, particularly when well rotted and in the form of dark brown crumbly humus, is very good at retaining water. In sandy soils it helps to fill the large gaps between particles, reducing the likelihood of roots being stranded in air spaces; in clay soils it helps to break up any compaction between the tiny particles and improves drainage whilst still retaining water.

Nitrogen in the soil

Most elements follow a fairly straightforward pattern of recycling – from plants, where they may be a part of large organic molecules, into the soil in dead material, broken down to simple mineral salts again, and back into plants. But nitrogen is an exception. It not only exists in the soil as soluble compounds of nitrogen with other elements (nitrate, nitrite and ammonium salts), but it is also present as nitrogen gas in air. It is cycled through these states and through protein in plants in quite a complex way. There are soil bacteria which break down dead protein material to ammonia and ammonium ions (the decomposers), and others which process the ammonium to give nitrites and nitrates. There are also specialized bacteria present in the roots of certain groups of plants which can use the nitrogen in the air to eventually make protein in the plants' cells. But there are also soil bacteria which convert nitrates back into nitrogen gas. On top of all this, electrical

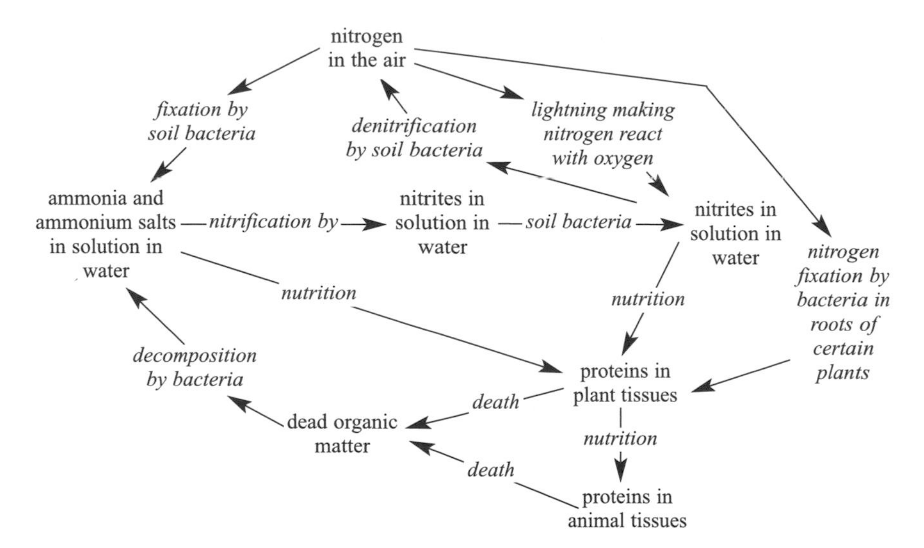

Figure 4-4. The cycling of nitrogen in the soil.

storms can convert nitrogen gas into ammonium. It's a profit-and-loss system which we can manipulate by adding yet more nitrate fertilizer and/or manure.

The relationships between the different processes can best be illustrated by Figure 4-4. No doubt you have recollections of being asked to learn the 'nitrogen cycle' at some stage during your school days. See my version on page 75!

There are further considerations which are of importance to gardeners. One is to do with the application of inadequately rotted manure or compost. If raw organic material, or plant waste which has only just started to decompose, is added to the soil, bacteria and fungi (see decomposition, Figure 4-4) will immediately get to work and reproduce in large numbers – it is, for them, a source of food. However, in fresh, leafy plant material there is a very high proportion of cellulose, which is a carbohydrate and contains no nitrogen. Protein is in relatively short supply. But bacteria *need protein*, just as we all do; so they turn to *other material* in the soil to get the necessary nitrogen. To begin with, then, it is quite possible that, after applying fresh manure or compost, you will actually *lose* nitrogen from the soil, stolen by bacteria and stored in their cells until they themselves die and decompose. This is why it is not a good idea to put your fresh vegetable peelings straight onto the soil.

How does this tie up with the practice, popular with organic gardeners, of using 'green manures'? The term usually refers to plants which have nitrogen-fixing bacteria in their roots and can make use of the nitrogen from the air (see fixation by bacteria, Figure 4-4). There is probably some sense in the practice, if, as is likely, these plants are particularly rich in nitrogenous compounds; the proportion of amino acids and proteins to cellulose and sugars is likely to be higher than in most plants, and certainly than in potato peelings and cast-off bits of cabbage and lettuce. However, I would suggest that green manuring is a system which warrants some rigorous scientific research to see whether, in fact, the decomposing material does indeed contribute to the available nitrogen in the soil, or whether the frantic bacterial activity during decomposition temporarily removes more than it gives out.

Recent research has apparently shown that there are certain plants with nitrogen-fixing bacteria in their roots which can directly convert their nitrogenous compounds into soil nitrites and nitrates. This would be extremely useful if reliable and needs to be further investigated.

Earthworms

I've already touched on the importance of earthworms in the section on managing soil structure. They are the gardener's good friends for many reasons. The presence of oxygen in the soil is vital. Molecules of oxygen are plentiful above ground, but for them to diffuse down through the gaps between soil particles is not so easy – particularly if many of those gaps are filled with water (oxygen moves much more slowly through water than through air). So the burrows made by worms are wonderful channels for oxygen to reach the roots of plants.

The worms also play an important part in the recycling of organic matter – pulling dead leaves down to consume in their burrows, and eating any other suitable material in the soil, such as bits of dead root. Take a good look at areas where there are leaves lying on the ground in autumn and notice if there are leaf stalks sticking up in the air. If there are it is a sign that earthworms have been active – they pull leaves down into their burrows and the stalk is often left above ground. Their food is digested by enzymes but not all valuable substances are absorbed, and their faeces are important contributors to the nutrient content of the soil. The great 19th-century biologist, Charles Darwin, is famous for his theory of evolution by natural selection. Less well known is his research into the activities of earthworms. His work was published in a book, delightfully entitled *The Formation of Vegetable Mould Through the Action of Worms with Observations on their Habits*. He laboriously collected and weighed worm casts and worked out that, in a field near his home in Kent, England, the worms were processing as much as 4.5 kg of soil per square metre per year. The figures in his original book are, of course, in pounds and square yards – the conversion is mine. His results have since been confirmed by other methods. If you have access to scientific journals you might be amused to follow up a paper I wrote, comparing the results Darwin achieved by means of primitive but painstaking methods with those worked out by a sophisticated computer-based method (*In Defence of Darwin* – see Further Reading).

Gardening suggestions

Making compost

There are so many guides to compost making that I am simply going to refer you to some good texts in the reading list at the end of the book. A strong recommendation, though, is to start a wormery (see page 78).

Mulching

Spread a good thick layer (5–10 cm) of compost on beds in late winter. This will gradually become incorporated into the soil by earthworms. It also helps to suppress weed growth at the beginning of the season. It's important to apply your mulch on to damp soil. It will then do a good job of retaining moisture in the event of a dry spring.

If you put it down when the soil is dry it will do an equally good job of keeping the soil dry! If you can stand the sight of black plastic use it as a mulch. We kept a vegetable patch moist and weed-free by covering it when rotating or when not needed.

Planting preparation

There are two conflicting schools of thought about placing compost in the bottom of holes dug for new planting. It has been customary to use this method for new herbaceous plants, shrubs or trees. There is now a view, though, which suggests that this is not a good idea for shrubs and trees because, as the compost decomposes, the soil sinks and leaves the plant in a depression which can collect too much water. Here is an opportunity for a scientific experiment, comparing one method with another - but to have any validity it would have to be done on a large scale with large numbers of plants and in different soil types.

Wormeries

If you haven't much space for a compost heap, or even in addition to standard compost making, it's really worthwhile establishing a wormery. Besides producing good-quality compost worms produce a fluid (worm 'pee') which can be diluted and used as a liquid fertilizer. They live in a plastic bin above a space for collecting the fluid, complete with a tap for drawing the liquid off. They can be fed most vegetable peelings and scraps (omitting citrus fruits and other acid materials) and leafy garden waste. In winter they need to be protected from cold, and it is important to prevent the contents of the bin becoming acid by adding slow-release alkaline material obtainable from wormery suppliers (see address at the end of the book). Otherwise they are trouble-free (and feeding them with vegetable peelings is a source of delight and amusement for my grandchildren).

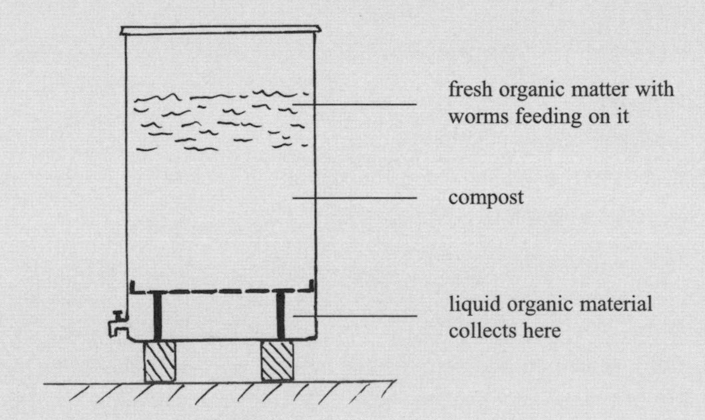

Figure 4-5. Sectional diagram of a typical wormery.

Is it possible to rely entirely on recycling?

In a natural woodland (tropical forest is an example *par excellence*) everything is recycled and the whole system sustains itself. So why not apply this principle to a garden? Well, in theory, it should be perfectly possible; in practice there are several snags. If you grow a lot of vegetables you are taking a whole crop out and probably putting only the peelings back onto the compost heap. Another problem may be that if you garden on sandy soil and perhaps, too, on a slope, rain water may leach mineral nutrients out and away from your garden. In a small patch, lack of space may prohibit you from composting absolutely everything and, in any case, woody material is not easily dealt with, even in a large garden. It is inevitable then that nutrients will be lost from your soil, albeit perhaps very slowly. Some 'feeding', with material obtained from elsewhere, is usually necessary.

Acidity and alkalinity – pH

The natural environments in which plants have evolved possess a great variety of soils, some of which are acid (boggy, peaty or sandy places), some more or less neutral (many clays and loams), some alkaline (on chalk or limestone). Our garden plants, therefore, vary in their adaptations to these different conditions. The term pH refers to a scale on which acidity or alkalinity is measured, 7 being neutral, acid less than 7, and alkaline more. Though plants have adapted to many different environments most chemical processes inside living organisms function best at a pH of nearly neutral, and most plants are happy in these conditions. Unless you wish to specialize in acid-lovers (calcifuges) or lime-lovers (calcicoles) aim for 6.5 to 7-ish.

The chemical form and availability to plants of many mineral elements is affected by pH. Nitrogen, phosphorus and sulphur are readily absorbed at the 6.5 to 7 level of pH. Calcium, magnesium and potassium are adequately available at this level too, though become more so as the soil becomes more alkaline. The real problem element is iron, which cannot be absorbed easily at pH values of more than 6.

Gardening suggestions

Soil-testing kits are available from garden centres and come with explanatory material. It's a good idea to take several samples from different parts of the garden to avoid a 'freak' result. It's also sensible to dig down a few inches below the surface which is where most of the roots will be. If your soil-testing kit involves making a solution of the colour indicator with water make sure you use *distilled* water (obtainable from chemists and garages) which should have a *neutral* pH. Tap water isn't always neutral.

Use of lime to neutralize acid soils If your soil turns out to be acid, and you wish to alter this, lime can be applied. Use hydrated lime (calcium hydroxide) or crushed limestone (calcium carbonate). Be sure, though, to follow the instructions on the packet – too much can be harmful. There is no point in liming unless it is absolutely necessary.

Use of peat to acidify alkaline soils, for example, for ericaceous plants. There are good reasons for avoiding the use of peat, if at all possible. The extraction of peat from natural habitats is causing a lot of harm; the conservation of peatlands is discussed in Chapter 6 (see page 125). If you garden on chalk or limestone and you have a conscience about conservation you might have to give up the idea of growing acid-loving plants!

The combined approach to encouraging good growth

The best approach is to use common sense, guided by the sort of basic scientific knowledge which I've provided, and beware of being swayed by proselytizing fundamentalists! If you have a ready supply of organic fertilizers of different kinds, and time and space to make compost, thank your lucky stars and use them. But if not, and you suspect that your soil lacks some minerals (plants not growing well or looking discoloured), don't hesitate to use a packet or bottle of man-made stuff from the garden centre. But follow instructions and don't overdo it.

5. Nature's arms race and how to negotiate a peace deal

Neither 'mother nature' nor 'nature red in tooth and claw' are really appropriate epithets for the non-human world. 'Nature's arms race', in which gardeners and farmers become intimately involved, is nearer to the mark. I suppose motherly qualities come to mind because all *our* nutritional needs have ultimately been met from wild plants and animals. But nature is not in the least motherly to its own offspring. Out there, be it in wilderness or garden, it's 'each one for itself'. This, of course, leads to the 'red in tooth and claw' concept but, curiously, things are not as bad as they may appear.

Ecological research has shown that individual species tend to occupy their own little niches, their particular way of 'making a living', to which they are well adapted, and in which they can avoid direct competition with other species for scarce resources. For example, several species of bird live on coastal mudflats, each endowed with a particular length of beak adapted to feeding on creatures found at different depths in the mud. They can thus all live happily together without having to squabble over their next meal. Admittedly the creatures living in the mud are being preyed upon but, at least, the birds aren't fighting each other for access to food.

Plants, like animals, occupy their own special niches – adapted to particular conditions. They are *ostensibly* docile and peaceable, quietly minding their own business. They just get on with making their own food, and are fortunate in their ability to survive nibbling by herbivorous animals. But many of them are remarkably good at fending off potential enemies – you'll have noticed that there are certain plants which slugs won't touch. Some of them are not at all innocent, entering the fray with a vengeance. There are some pretty nasty weapons in their armoury, stinging cells, spines, alkaloid poisons to name but a few. So, if 'red in tooth and claw' conjures up snarling carnivores, nature could equally well be described as 'green in thorn and killer chemicals'! There is thus a constant arms race going on – organisms needing to feed, but also having to avoid being preyed upon or outwitted in competition. We, as gardeners, are very much part of that arms race because we are in the business of manipulating things to suit our needs.

It makes sense to think of the living world, the biosphere, as an intricate and miraculously self-sustaining network. The expression 'the balance of nature' is sometimes used in connection with this network, but it's not necessarily a balance in the sense of a see-saw which is poised, equally weighted at either end. It is, rather, a constantly fluctuating balance – a see-saw being played on and moving up and down. Sometimes, due to some cataclysmic event or to human mismanagement, the fluctuations can get out of hand, drastic crashes occur (and the players may fall off!). Humans have not treated the biosphere well, so gardeners, being intimately involved in the network, have a particular responsibility to be constructive and careful in their management of it. This chapter is about being involved in the arms race but achieving a sensible balance without being too destructive.

Because plants and animals, and their relationships with their environment, are so complex, the science of the topics in this chapter is by no means precise. What follows is, I hope, a guide through a somewhat tortuous maze.

Food webs

Ecologists – who study living things in relation to their environment – have tried to work out the way in which species fit into their niches and to unravel the complex interactions between them. Feeding relationships are important here. Who feeds on what can be determined by direct observation, or by analysing the contents of a predator's digestive system. A picture emerges of *networks* of feeding relationships, referred to as **food webs**. An example of a garden food web might be depicted as in Figure 5-1. It must be emphasized that most diagrams like this are gross over-simplifications. There are more creatures in our gardens than we might suppose (see Chapter 6), and finding out all their feeding habits would be a challenge indeed.

Weeds, pests and diseases

In the food web in Figure 5-1 there are plants to be cherished, a plant which is not welcome, animals which can be tolerated, animals which are a nuisance, and animals which are positively beneficial. This is just a very small selection of the kinds of creatures and their feeding relationships which *might* exist in a garden. There is also a myriad of micro-organisms, busy being parasites, often causing nasty diseases; some, on the other hand, are going about their business harmlessly, and some are being really helpful. Organisms which we call pests or weeds are simply those which militate against our efforts to produce beautiful flowers, immaculate foliage, and copious, unblemished vegetables. In Figure 5-1 they are

Figure 5-1. A hypothetical garden food web.

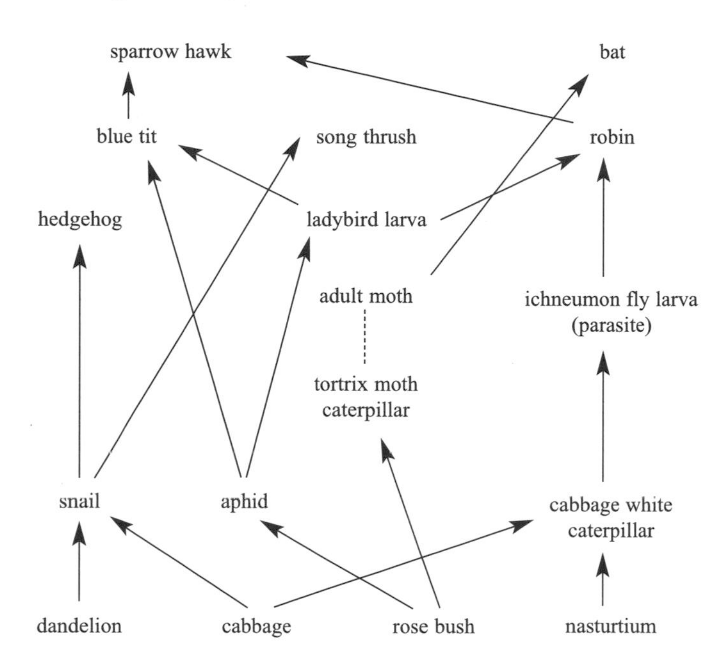

the cabbage white butterfly caterpillar, the tortrix moth caterpillar, the aphid, the snail and the dandelion.

There is no point in listing the numerous different kinds of nuisances and ways of identifying them. There are excellent books available and you will find some suggestions in the Further Reading list. If they are not obviously available on your library shelves, or in the local garden centre, remember that libraries will order things for a small fee.

Preserving the cherished, the tolerable and the beneficial, whilst keeping at bay the undesirables

There is a school of thought which insists that it must all be done 'organically', relying on natural predators and parasites, using hand or mechanical removal of weeds or other nasties, and spraying only with substances which occur naturally or are approved by the Soil Association. Others think this is a load of nonsense – a visit to the chemicals section of a garden centre demonstrates that there is a vast army of gardeners who spray all manner of man-made concoctions on their plants. Scientific evidence would suggest, though, that – as with soil management – the most sensible approach is to keep an open mind and adopt a combination of both.

The next few sections contain guidelines towards achieving a *compromise* – the use of a combination of biological and man-made chemical agents. This kind of approach is technically called **Integrated Pest Management (IPM)**. In general it makes sense to use the predators and parasites in natural food webs to keep things under control. This is particularly so in a garden where the scale of operation is fairly small. They are not, however, likely to eradicate a pest species completely, for reasons I shall explain. If things are really troublesome, modern science offers well-tested alternatives, though man-made pesticides are not without their problems – not least the fact that pests become resistant to them. Some people worry about the human health and safety aspects of pesticides but the Food and Environment Protection Act, 1985, the Pesticides Safety Directorate, and European legislation are there to safeguard us. In the USA there is the Federal Environmental Pesticide Control Act of 1972. All products legally permitted for use in agriculture and horticulture have been thoroughly tested for safety. It is therefore technically illegal to attack the aphids on your roses by spraying them with diluted washing-up liquid; detergents haven't been tested with their gardening safety in mind!

One of the themes of this book, though, is that science isn't perfect – scientists do the best they can. It's worth remembering that just as, in medicine, there may be changes of treatment in the light of new discoveries, so scientists working on pesticides may change their tune in view of new evidence. What was once thought safe may be deemed no longer so, and products are taken off the market (as has happened recently under European rules). The range of pesticides available to gardeners is now very limited. It's my view that the danger which this small selection poses to humans, domestic animals and the environment is utterly insignificant compared with other risks of modern life.

Controlling animal pests without artificial chemicals

This section covers those methods which are commonly termed biological, physical and cultural – those which rely on the use of other living organisms, traps and barriers, or particular cultivation practices.

Predators and parasites

There are many examples of predators or parasites which make their living by consuming undesirables in the garden. Blue tits, ladybirds and lacewing larvae eat aphids. There are tiny parasitic wasps which have the unpleasant habit of laying their eggs inside the larvae of moths or flies, so that the grubs consume their host from the inside – *Encarsia*, used in glasshouses to control white fly, is an example. But there are snags about counting on these agents. One factor is the relationship between predator and prey, a topic which has been much studied by ecologists.

Interactions between predators and prey

The relationship between a single predator and its prey often follows a particular pattern. At first the predator has an abundance of prey and is able to reproduce so successfully that its numbers increase in parallel with prey numbers. But there comes a point where predation begins to reduce the numbers of prey, there is less food available for the predator, and its numbers begin to decline. After a phase of reduction in predator numbers the prey species begins to recover and the pattern repeats itself.

Scientific evidence for this effect comes from situations where the ecosystem is a simple one. An example is a Canadian study, published as long ago as 1937, which collected and collated hunting and trapping records of the numbers of lynx (predator) and snowshoe hare (prey) caught over a period of almost 100 years. The results show a striking picture of approximately 10-year cycles; as numbers of hare increased so, after a brief delay, did the numbers of lynx, but there always came a point where hare numbers started to decrease followed by a crash in lynx numbers. Then hares recovered and so did the lynx. I won't bother you with the graph – but it shows a very pretty and obvious 'boom-and-bust' pattern, with the lines for lynx and hare running up and down in parallel with a slight lag of lynx behind hare. Another more recently studied example is that of lemmings and their predators on the arctic tundra of northern Europe. Here lemmings are the main

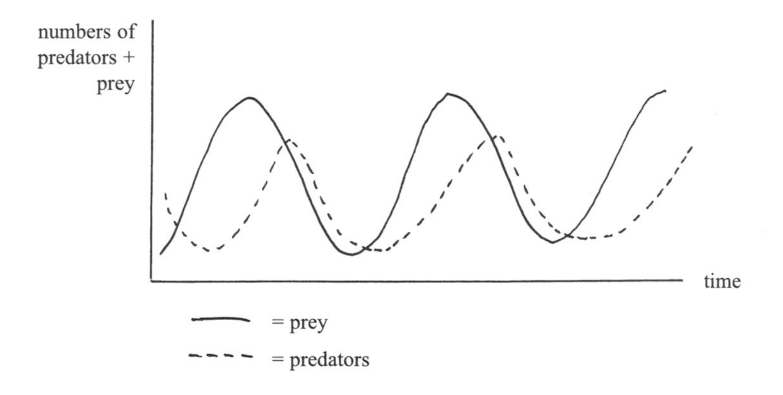

Figure 5-2. The kind of relationship which often exists between numbers of predators and their prey.

small herbivores, and the carnivores which prey on them are arctic fox, snowy owl, long-tailed skua and stoat. Lemming numbers have long been known to fluctuate markedly, leading to various myths, including mass suicide, and hypotheses to do with weather and availability of food plants. Over a 15-year period these animals have been closely studied, and the 'boom-and-bust' story has been shown to hold good for them too. As lemming numbers increase so do the numbers of predators, reaching a point where the pressure on the lemmings is such that their numbers start to decrease. With not enough food around the foxes, owls and skuas apparently move off to search for other prey species which, in the arctic, are not numerous, and the stoats, which feed only on lemmings, die off. The lemmings then begin to do better and the cycle starts again.

In a garden similar cycles may occur. An example is the potential interaction between aphids and their predators. Ladybird larvae might come across an abundant supply of food on your rose bushes. A veritable feast ensues and the ladybirds do so well that they multiply prodigiously. Then there comes a point when aphids are beginning to be in short supply and the ladybirds can no longer support so many offspring. But as soon as predator numbers die down, the aphids begin to have a field day again and return to their former abundance. So you are back to square one.

Parasites

The story is simple. To be a successful parasite you mustn't kill your source of food straight away. A parasite is an organism which feeds from a *living* plant or animal – its host. Much of the time the thieving beast sits happily consuming nice nutritious juices from its host's body, occasionally producing vast numbers of offspring, eggs, seeds or spores, which are specially designed to invade a new host. The crucial thing is that a *successful* parasite doesn't immediately finish off its host, thereby running the risk of being thrown out of its cushy lodgings before having had a chance to reproduce or to emerge as an adult. The host is simply weakened. So, while parasites may be a useful way of reducing the virulence of a pest species, and perhaps reducing its capacity to multiply, they will rarely wipe out the pest altogether.

Gardening suggestions

Predators and parasites in the garden

By all means encourage predatory species – ladybirds, lacewings, ground beetles, insectivorous birds – but don't expect them to do more than periodically keep the numbers of pests at moderate levels. Lacewings can be encouraged by providing them with overwintering quarters – such 'lacewing hotels' can be purchased now through gardening catalogues. Ladybirds often hibernate in hollow stems, so leave tidying away some of the stems of large herbaceous perennials until warm weather in the spring. Nematode worms which parasitize slugs can be purchased by mail order from gardening catalogues, particularly from firms specializing in 'organic' gardening.

Greenhouses and conservatories

The Victorians used to introduce frogs into their greenhouses and encourage them to stay there. Why not try it? Use such parasitic organisms as are on offer for use in glasshouses (obtainable from catalogues) but don't expect too much of them. Examples are the parasitic wasp, *Encarsia*, for controlling greenhouse white fly, nematode worms which transmit fatal bacterial diseases to vine weevil grubs, and the toxin-producing bacterium *Bacillus thuringiensis* for moth and butterfly caterpillars. Don't sweep away spiders' webs; spiders are the tigers of your jungle and will gobble up any pests which happen to land in their well-laid traps.

Safety in numbers (of different species)

Where a food web is a simple one (as is the case with the lemmings and their predators) there is the danger of sudden surges in numbers – the danger of plagues of pests. If, however, the situation is much more complex, with both herbivores and carnivores varying their menu so that as soon as one item begins to decline another is chosen, plagues of any one species usually don't occur. This tends to be the case in the more species-diverse communities of temperate, Mediterranean and tropical climates than in the simpler ecosystems of polar regions. Before humans invented agriculture, starting to grow large areas of a single crop, major fluctuations in numbers of animals were probably rare. Locusts, for example, would once, perhaps, have existed in relatively small populations, feeding on a variety of plants and switching from one species to another as the need arose but never finding a cornucopia of especially nutritious food all in one place. An example, about which we know a good deal, is the case of the Colorado beetle. In 1824 an insect collector exploring the Rocky Mountains discovered a few scattered specimens of a striped beetle feeding on plants belonging to the potato family. He named it as new to science. Then 30 years later settlers brought the cultivated potato and planted them as a crop in that region. The beetle proceeded to take advantage of the copious food supplies, spreading eastwards as a major pest and cause of famine, subsequently reaching Europe where it was greatly feared. We don't hear much about it now, but crop scientists are constantly on the alert for its reappearance.

There is little concrete evidence for the plague effect in gardens, probably because gardens are small-scale compared with agricultural fields where there may be hectares of a single crop. But it makes sense to bear it in mind as a principle on which to base gardening practice.

Gardening suggestion

It is probable (though not inevitable) that the more complex your garden food web, the more likely it is that pests will remain at relatively harmless levels. Try to avoid large areas of a single species (vegetables can be intercropped with other plants - see companion planting, page 94), plant thickly and, again, encourage birds, frogs, hedgehogs and so on (see Chapter 6 on encouraging wildlife).

Pheromones

These are chemical substances, produced by many insects for the purpose of communicating with one another. They act as a sort of insect language. One type of pheromone is a sexual attractant and male moths, for example, can detect such a chemical given off by a female at enormous distances away. As soon as this was discovered it was obvious that pheromones could be used in traps. Sticky traps impregnated with synthesized sex pheromones are used to trap male moths. They are used in conjunction with conventional pesticides – an increase in the numbers of male moths caught indicates the peak time for applying the insecticide and preventing the females mating and laying eggs. This is a good example of the **integration** of a biological method with a conventional pesticide.

Traps, barriers and murder

So far I haven't mentioned the dreaded words – slugs and snails. If one were to do a country-wide poll of gardeners, asking which they find the most troublesome garden pest, I wouldn't be surprised if molluscs came out top. Rabbits and pigeons might come a close second. Preventing these voracious foragers getting to their food is often the solution, whilst luring them into traps or outright murder may be the last resort.

Gardening suggestions

An obvious way of preventing persistent nibbling is to put up a barrier - fencing for rabbits, penetrating a considerable way into the ground; netting in the case of birds; fleece for carrot root fly. Mini-electric fences have been tried for slugs, as has copper wire and tape. Barriers can be made with crushed eggshell, sand or cinders which slugs are reputed not to like crawling over.

In my experience the barrier method for molluscs isn't particularly effective. But you could design an experiment to test what might work best for *your* slugs and snails. Collect a few and place them inside circles of various materials and watch to see whether they get out, and if so, how quickly! Watch, also, to see which plants they seem to like best and concentrate on protecting *them*. I have found, I fear, that nothing really protects those plants to which slugs and snails are particularly partial (in my garden – Michaelmas daisies, *Rudbeckia* spp. and *Ligularia* spp. – oh! and delphiniums, which I have given up altogether).

If you don't mind killing, beer traps are a good idea and effective in reducing the population somewhat. I've collected as many as 150 slugs per night in four or five such traps; but they continue to go marching on. I have to confess that on damp, mild nights I can be seen patrolling the garden with a torch and a pair of scissors. (What I commit with the scissors I leave you to guess.) I hope I don't offend those of a sensitive disposition but, on the same note, manual removal and squashing of caterpillars and clusters of aphids is a useful line of attack. Don't forget to look on the *undersides* of leaves. If you were a nibbler, where would you hide? (See Chapter 3, page 48 for a way of protecting trays of seedlings from slugs.)

Reams have been written about the slug problem. A delightful little manual, which I recommend, is *The Little Slug Book*, produced by the Centre for Alternative Technology (see Further Reading).

Biological control of weeds

Since a weed is a plant which you don't want in your garden (usually because it competes with those you *do* want) one way of approaching the problem is to make it impossible for weeds to find any *space* to grow – not so easy in practice (see Gardening suggestion overleaf).

Competition

Despite the fact that many species avoid competing with each other by occupying a special ecological niche, there still exists a certain amount of competition between species (ecologists call it **interspecific competition**), particularly amongst plants, all jostling for the same resources: light from above and water and minerals from below.

Gardening suggestions

Mulching

Cover any bare ground as thickly as possible with a layer of some suitable material such as manure, chopped bark, old carpet or porous plastic membrane.

Thick planting

Try to grow your desirable plants as close together as possible – a recommendation often made in gardening books and magazines. There is a practical problem here, though, not often mentioned. The situation in winter is different from that in summer. It is difficult not to leave some bare ground in winter, when your herbaceous border has died down, or when you have cleared away the annuals (unless you are extremely assiduous and cover it with a mulch immediately!). During the mild winters we have recently been having new weeds start to get a foothold in autumn and early winter. I have that ubiquitous bitter cress flowering in midwinter; once seeded it sows trouble for the spring and summer! So try for thick planting in summer, combined with speedy mulching in winter.

One year's seed – seven years' weed

So true! Be watchful and destroy your weeds before they seed.

Learning to love weeds

Another approach is to learn to tolerate weeds – even to garden with them, as is done in a French garden in the Loire valley (Ferme du Chateau, Chaumont-sur-Loire). A favourite poem of mine is one by Norman Nicholson on just this theme. Here is part of it.

Some people are flower lovers.
I'm a weed lover.

Weeds don't need planting in well-drained soil
They don't ask for fertilizer or bits of rag to scare away birds.
They come without invitation
And they don't take the hint when you want them to go

Even the names are a folk-song:
Fat hen, rat's tail, cat's ear, old men's baccy and stinking Billy
Ring a prettier chime for me than honeysuckle or jasmine
And Sweet Cicely smells cleaner than Sweet William though she's barred
from the garden.

You can keep your flowers
Give me weeds.

(From Norman Nicholson, *Collected Poems*, Faber & Faber, 1994, by permission of David Higham Associates)

Once, as we were walking round a field next to the garden of a student of mine, he pointed out a little yellow flower. He asked me what it was and proceeded to pour scorn on it for the nuisance it caused him. This was the lesser celandine, *Ranunculus ficaria*, one of my very favourite wildflowers because of its shiny, cheerful face in spring. It does spread rapidly, and has little underground bulbils which break off as you weed, but I wouldn't be without it – its habits just have to be tolerated. A vigilant eye though, and a thorough weeding wherever it pops up in a particularly undesirable place, are a good idea. I have to admit that if one inherited a garden where it was rife the task of controlling it would be difficult. There are, in fact, cultivars available. The Royal Horticultural Society's *Plantfinder* gives comprehensive information about which nurseries supply which plants and whether any less invasive varieties are available.

If you decide to try 'gardening with weeds' it is perhaps courteous to think first of the neighbours! A pretty patch of rose-bay willow herb is *one* thing, but the wind-borne seeds it produces will hardly endear you to friends next door. The jolly yellow flowers of ragwort could look good in a border too, but it also produces copious wind-dispersed seeds and, moreover, is a registered agricultural weed which can kill cattle and horses. So there are hazards. Relax, but use some common sense.

Diseases

The causes of disease in plants are very similar to those causing many of our own illnesses – invasion by viruses, bacteria, fungi or worms. As we are taught for our own good health, hygiene is an important key to keeping these attacks at bay. It's a fact of life that the fungus causing athletes' foot flourishes in the damp and ill-ventilated regions between our toes. The same applies to plants – for example, the grey mould (*Botrytis cinerea*) flourishes in the moist and still centres of indoor plants. An exception is the powdery mildew which, ironically, is particularly troublesome after hot dry spells. Once a fungus produces its myriad reproductive spores it will spread rapidly. The spores settle on the surface of leaves, stems or roots and germinate into a mass of fine threads, the hyphae, which penetrate the plant tissues, in much the same way as do the beneficial fungi described in Chapter 2 (see Figure 2-3, page 23).

Recent research has concentrated on beneficial fungi, and it may well be that in the future we shall be able to manipulate things with the aid of these. Look out in the next few years for practical applications.

Nowadays one of the hazards is the enormous traffic in exotic plants from abroad. These bring with them diseases, and although they may be specific to those plants they may on the other hand spread to other species. A strict watch needs to be kept.

Bacteria and viruses are spread by sucking insects (vectors), as is the case in our own devastating illnesses such as malaria. Another particularly nasty one is Lyme disease, identified as recently as 1975. It is transmitted by bacteria transmitted to the human bloodstream by the bite of various species of ticks. It can be very serious, so keep your legs, and those of your children, protected when walking through long grass in the countryside

Getting rid of the vectors is one of the preventive strategies. Nematode worms burrow their way into tissues as in bilharzia and other tropical diseases. The worms must therefore be prevented from multiplying in their habitat or, often, in another host organism.

Detailed descriptions of plant diseases is beyond the scope of this book, but there are good guides available (see Further Reading).

Gardening suggestions

Vigilance and hygiene are the keys. If diseased leaves, or the very first aphids to appear, are removed immediately, the potential for the spread of spores or the introduction of disease organisms into tissues is considerably reduced. Likewise plants affected by viruses (mottled or streaked leaves are signs) should be destroyed because sucking insects may spread the disease to other plants. Don't forget, also, that pests can lurk on the undersides of leaves, so look there as well. Fungal spores can hang around for a long time, so diseased leaves on the ground should be cleared up and preferably burnt. The clearing up of black-spot-affected rose leaves is particularly important, for example. Composting is alright only if your compost heap really does heat up to 50°C. In the greenhouse try to start the season with clean plant pots.

Rotation

Soil hygiene in agriculture is helped by rotating crops so that there is a period when disease organisms do not have their favoured host available and so can't multiply. Having said that, gardens and allotments tend to be so close together, and the areas of plants so small, that spores and worm eggs can easily be transferred on human or bird feet, tools, wheelbarrows, and so on from one patch to another. Crop rotation is therefore not as effective on a garden scale as it is on the larger agricultural scale.

General good health

These days we are constantly being urged to improve our resistance to ill health by good diet and exercise. We're not required to take our plants for walks, but we can make sure that they have optimum nutritional and growing conditions. A healthy plant, with a good strong cuticle covering its leaves, strong-walled conducting vessels and foliage well spread out for maximum light collection and ventilation, will be less susceptible to attack by micro-organisms.

Plant resistance to disease

This is where the 'arms race' analogy is very relevant. Many micro-organisms make their living by muscling in on the food or enzyme mechanisms available in plant cells. If this causes serious harm and decreases a plant's ability to reproduce, any change in the plant tissues which counteracts the effects of the invader will be beneficial and the genes which control that characteristic will be passed on to the next generation. For example there might have been a genetic change (see mutation, Chapter 7, page 132) which confers on them the ability to produce a chemical substance which destroys the disease organism or which deters a vector (carrier) from nibbling or sucking. Numerous such 'strategies' occur in the wild, particularly amongst tropical plants, many of which produce the most alarming toxins to deter their enemies. Here we have the arms race operating with a vengeance – plants producing substances or structures to protect themselves, their enemies evolving mechanisms for circumventing these defences, and the plants being under pressure to develop new survival strategies.

A great deal of horticultural research has been undertaken to find plant varieties which have resistance to disease. It is a never-ending task, because as soon as a suitable variety of one species is found another may succumb to a new ailment. For example, in the UK no sooner has Dutch elm disease, caused by the fungus *Ophiostoma ulmi* (synonym *Ceratocystis ulmi*), been controlled than sudden oak death, *Phytphthora ramorum*, becomes a new threat.

One of the most powerful arguments for wildlife conservation is our need for as varied as possible a genetic base from which to breed, or genetically modify, our cultivated plants. Likewise we need to continue to conserve some of the more unusual varieties of all our garden plants – hence the importance of national collections of species. Whilst we have been busy breeding for attractive colour and shape of flowers and foliage, and productivity and uniformity in our vegetables and fruit, we have tended to ignore disease resistance along the way. But the wild relatives from which all our cultivars have been originally derived often possess genes which confer resistance to disease. Looking to them in the future could be very useful.

Gardening suggestions

Purchasing resistant varieties

Gardening catalogues from conventional commercial nurseries often mention varieties which are resistant to particular problems. Because the proponents of 'organic gardening' don't permit artificial chemical control of diseases, organic horticulturists have specialized in promoting resistant strains; organic gardening catalogues are therefore particularly good sources. There are also ways of legally acquiring varieties which are not on the European list of permitted saleable items. By becoming a member of a club and receiving complimentary samples one can, for example, acquire unusual vegetables. I suggest making enquiries of the Henry Doubleday Research Association (see Useful Addresses).

Companion planting – does it work?

There is a great deal of folklore about plants which fend off pests and are useful to vulnerable neighbours, and much has been written on this subject. In my experience lists of companion plants, and the help they are supposed to give, tend to be somewhat vague. They differ wildly from each other and are sometimes even contradictory. In one particular newspaper article, for example – 'Marigolds protect most plants from insects and slugs' – what kind of marigolds? pot marigold, *Calendula*, or African marigold, *Tagetes*? Again, in the same article '...summer savory keeps the bean beetle away from beans', but in an organic gardening book I also read that summer savory should be used as a precaution against black fly. Perhaps it does both. *Tagetes*, the African marigold, is reputed to have a 'powerful effect' on ground elder and to 'kill it quite effectively'. I also note that nasturtiums are supposed to 'deter' aphids, cabbage white butterfly and fruit pests from tomatoes, radishes, cabbage and cucumber. I presume this means that these pests like it so much that they don't bother with other plants. I do know that cabbage white caterpillars love nasturtiums but I would have thought that, if planted next to precious crops, the nasturtiums might equally well flag up, to all such pests in the neighbourhood, the location of delicious cabbages and cucumbers. There is one recommendation which does appear consistently: grow garlic under your roses to ward off aphids. That is one which is reputed to be genuinely effective, though I'm not sure whether a balmy summer's evening wander down the garden to smell the roses would be enhanced by the accompanying odour of garlic. At the time of writing there have been reports that a garlic extract is being developed as a commercial pesticide. So it looks as though there is genuine value in growing it as a pest deterrent.

I am sceptical about companion planting, but all these things are worth trying and there is ample scope for your own scientific experiments.

On a more serious note, trials which *have* been carried out to test schemes of companion planting have been inconclusive. There is, however, scientific evidence that some plants produce a particular substance called methyl jasmonate which is known to play a role in the chemical processes by which plants can put up some resistance to pests. *Artemisia* (often mentioned in lists of companion plants) produces large amounts of it and it is volatile, so could well reach neighbouring plants and help them to fend off harm. The African marigold and its chrysanthemum relatives produce pyrethrum which protects them against insect damage. Perhaps insects *do* avoid areas where there are patches of these plants.

Successes and failures of biological control

Much of the research on biological pest control has been done in agriculture, and there have been some spectacular successes. One of the most famous was the rescue of the Californian citrus fruit industry after the cottony cushion scale insect had virtually annihilated it in the late 1880s. The pest originated in Australia, so an idea was mooted to import one of its natural predators from 'down under'. An Australian ladybird was introduced to the orchards, and control of the pest achieved in less than two years. The success was maintained – possibly because the ladybirds were able to find other food when the scale insect numbers were reduced, and so were able to keep their numbers up (though I have been unable to substantiate this). Another success story was the classic case of the control of prickly pear – a cactus-like plant which was becoming a serious weed in Australia. The solution was achieved by introducing a moth whose caterpillars ate the plant.

There have been some notorious failures, though. In Puerto Rico a ladybird predator was introduced to control mulberry scale insect. At first this was successful and the scale insects virtually disappeared. But 17 years later a sudden heavy infestation occurred – an example of the 'boom-and-bust' effect described earlier. Ecologists thus learnt that it might be necessary to provide alternative food – a more complex food web – to tide the predator over in times of prey scarcity.

The moral of all this is that there is no easy 'quick fix', and the most sensible approach is to keep an open mind, know as much as possible about the ecology of your plants and animals, and adopt a combination of different methods.

Using man-made chemicals

Some definitions might be helpful:

- **Herbicide** – anything which kills plants. More commonly called weedkillers they may be able to kill most plants or they may be selective, that is, designed to kill only certain kinds of plant: for example, only broad-leaved species (dicotyledons), only those with parallel-veined leaves such as grasses and cereals (monocotyledons) or only ferns (for example, asulam, which was designed specially to kill bracken).
- **Pesticide** – a general term meaning a substance which will kill any kind of pest, usually referring to animal pests.
- **Broad spectrum** – pesticides designed to kill a wide range of organisms.
- **Selective** – a pesticide designed to kill a specific type of plant.
- **Insecticide** – fairly obviously a substance which kills insects.
- **Molluscicide** – deals with slugs and snails
- **Vermicide** – kills worms (usually earthworms).
- **Nematicide** – kills nematode worms.
- **Systemic** – means that the substance, when applied to a plant, becomes absorbed into the plant's tissues.
- **Contact** – the substance acts simply by contact with the surface of plant or pest.
- **Persistent** – refers to the fact that a compound remains in the environment for a long time, if not forever.
- **Biodegradable** – means capable of being broken down (usually by the action of micro-organisms).

In the years following World War II, when the UK was desperate to become more self sufficient in agriculture, there was a major drive towards developing and marketing the chemical weapons of mass pest destruction which had been researched in the 1930s. The first to be used on a large scale were the organochlorine insecticides, DDT, aldrin and dieldrin, followed by organophosphates and carbamates. All these are neuro-toxins – they kill insects by damaging their nervous systems. Because our nerve cells function in a similar way to those of insects these chemicals do have *potential* to do us harm if exposed to large doses. There were – and still are in developing countries where regulations are less stringent – cases of farmers becoming ill with neurological symptoms. The organochlorines were also demonstrated to be responsible for the deaths of birds of prey, accumulating, as they did, in fat deposits in animals and building up to high concentrations in the predators at the top of food chains. In the 1960s there was a big backlash, fuelled by an influential book, *Silent Spring,* by the American biologist, Rachel Carson. This stimulated research into more benign substances. Modern pesticides *are* less harmful to humans, other mammals and the environment. Examples are the insecticides based on pyrethrum-like compounds (pyrethrum is a chemical found in African marigolds and its relatives). It's sometimes not realized, too, that not all pesticides build up in food chains – only the ones which tend to be fat soluble and remain in the body, stored in fatty tissues. So not all modern man-made pesticides need to be treated with fear and loathing.

Vigilance and caution are still necessary. Permethrin, a synthetic compound related to the naturally occurring pyrethrum, is such a powerful killer of insects that it can wreak havoc if it attacks the wrong targets. It may kill beneficial predators or, if released into water courses, decimates the insects on which fish depend for food.

Herbicides, such as glyphosate and paraquat, contain compounds which interfere with the manufacture of amino acids (precursors of proteins) in the plant, or with the reactions involved in photosynthesis. Since the chemistry of these processes is very different from anything that goes on in mammal bodies they are considered to be less potentially harmful to us (though you would be very much the worse for wear if you swallowed a bottleful of Tumbleweed!). Research is proceeding into studying the effects of growth-regulating hormones on aspects to do with controlling weeds. This is an exciting field and promises to yield results in the future.

Pesticide resistance

A familiar human health problem these days is that of antibiotic-resistant bacteria which cause life-threatening disease. Unfortunately the same kind of difficulty rears its head with pests which become resistant to pesticides. A spontaneous genetic change may occur (see Chapter 7) which gives a pest the ability to digest the chemical used against it, or to break it down internally and eliminate the chemical from its body. The individuals able to do this will survive, reproduce and pass on this resistance to their offspring. All the others will, of course, be killed off. And so a new resistant strain becomes established. Greenhouse white fly is an example of an insect which has become resistant to many insecticides. Purveyors of pesticides therefore have to keep ahead of the game – the 'arms race' analogy is particularly pertinent in this context.

Gardening suggestion

Safety

If using man-made chemicals the message is simple - always follow religiously the instructions on the container. Find out, too, from any instructions with the product, or from the waste disposal department of your local authority, what to do about getting rid of chemicals you no longer require.

Common sense and a combined approach

As with much of science, our knowledge of the effects of pesticides is constantly being reassessed and modified. There has recently been there a complete overhaul of European regulations, and many products previously available to gardeners have been taken off the permitted list. The best I can do at the time of writing is to refer you to the government website (see Useful Addresses).

The science of both biological control and man-made chemical control is by no means precise, largely because plants and animals, and their relationships with their environment, are so complex and variable. As explained earlier, organisms are constantly interacting with each other, involved in their own 'arms race' in their 'struggle' to survive. We may enter the arena with a strategy which we think may be effective, only to find that we are outwitted. This is an area where your own observations, experiments, the adoption of the combined approach (IPM) and a good dollop of common sense really pay off.

Gardening suggestions

Spot treatment of weeds

Where you wish to get rid of troublesome weeds amongst already established flowers or vegetables, and hoeing or hand weeding isn't an option, then rather than spraying herbicide use a paint brush, glove or other wiping device. Bindweed can be controlled by encouraging it to climb up canes and then spot treating it in this way.

Care with fungicides

In Chapter 2 mention was made of the fungi which form beneficial associations with plant roots (mycorrhizae, see page 22). It's therefore essential to be careful about not spilling fungicide on the soil.

Live and let live – does a garden need to be immaculate?

Finally, why not give some thought to accepting a degree of imperfection in your garden? Any peace deal which we manage to negotiate is likely to be only temporary. Some rogue is bound to pop up somewhere and we have to be on the alert again. That is in the nature of the dynamics of life and the complex web of nature.

Adopt a positive attitude and remember that diseases or 'sports' sometimes turn out to be interesting. Take the virus diseases of tulips which produce the strange streaking of the petals called 'breaks'; the affected varieties eventually became very fashionable and desirable. The variegated foliage of so many of the plants we cultivate has often been the result of an unexpected genetic change which, in the wild, would have probably disadvantaged the plant, since it invariably resulted in less chlorophyll for photosynthesis. To put another positive slant on things – take a good magnifying glass and have a close look at an aphid or a white fly. Meditate for a moment on the miracle of design and biochemical engineering which it represents! Do we really want to eliminate these beautiful creatures from our world?

6. Biodiversity in the garden – encouraging wildlife

'Wildlife gardening', or 'gardening with nature', has become a popular topic in recent years. Articles appear in mainstream gardening magazines and there are several books currently in print. But the idea isn't entirely new; William Robinson wrote *The Wild Garden* in 1870. He was an eminent Victorian gardener and writer who introduced the idea of allowing native wild species to take up residence in our cultivated plots. Rather than attempting the purposeful conservation of threatened species, his concept was simply to allow his plants – exotic or otherwise – freedom to establish themselves where they chose. He was also not averse to exchanging wild plants with those from friends in distant parts of the country, something which would be frowned upon today.

The idea of actually developing gardens as nature reserves came to a wider public in the 1980s and 1990s, when conservation organizations began to promote wildlife gardening as a way of compensating for the disastrous losses of wild species in the countryside. There is now copious literature on the subject, but one of the earliest books of that pioneering period was *How to Make a Wildlife Garden* by Chris Baines, an energetic and enthusiastic promoter of the idea that nature conservation is possible in the gardens and open spaces of our towns and cities.

There is also scientific research to support the notion that gardens can be home to an enormous variety of species. In England an unassuming Leicestershire gardener and biologist, Jennifer Owen, carried out surveys of species in her fairly conventional suburban patch over a period of 15 years, and wrote *The Ecology of a Garden – the First Fifteen Years.* She recorded a total of around 2,200 species of plants and animals in her garden. Of the plants, many were deliberately planted cultivars, as would be expected; some were introduced native wild species and some simply self-sown. There were some animals, difficult to catch or identify, which she didn't study at all, so her estimate of just over 2,000 is certainly conservative. *The Natural History of a Garden*, by Colin and Geoffrey Spedding, gives another fascinating insight into the private lives of the inhabitants of a domestic garden; and recent research at Sheffield University, England, the Biodiversity in Urban Gardens (BUGS) Project, has revealed that almost any garden, whether or not deliberately designed for wildlife, can provide homes for wild species.

Before going further, a word about **biodiversity**. This somewhat daunting term has entered the vocabulary of biologists and wildlife conservationists in recent years and now crops up everywhere. It was originally coined by the great American biologist, Edward O. Wilson, who has been a champion of the need to save our planet from a major extinction of species. Biodiversity simply means the variety of life – that is variety of species, and of genetic diversity within species. Once a plant or animal becomes extinct some of that biodiversity is lost; if you think back to Chapter 1, this loss is ultimately a *chemical* one. This pinpoints one of the reasons why conserving living organisms is so important. With any species which becomes extinct we may, for instance, be losing valuable sources of useful pharmaceutical substances. Examples abound of compounds extracted from plants which have been used effectively in medicine – an extract of the rosy periwinkle (*Catharanthus rosea)* for the treatment of leukaemia, an anti-cancer drug from the Pacific yew (*Taxus brevifolia*), and, more recently, from the UK's native species (*Taxus baccata*). The most familiar is aspirin, derived from the salicylic acid found in willow species (*Salix* spp. – hence the term 'salicylic') and meadow sweet (*Filipendula ulmaria*)

In agriculture, too, entirely new breeds of plants or animals might well be needed to cope with changing climate. It would be wise, therefore, to keep as wide a genetic base as possible to provide the material from which to breed, or to use in genetic modification (see Chapter 7). Wildlife conservation therefore has utilitarian value. Biodiversity is vital too for people's spiritual and psychological welfare – a varied countryside is a lot more aesthetically pleasing than a boring monoculture. I can recommend two passages to bring this home. First, Charles Dickens' description of the joy and peace which Oliver Twist gained from wildflowers and the countryside after his traumatic experiences with the criminal gang. Then Katharine Swift's revelation – 'Gazing into the cow parsley was like looking up into the milky way…all the panic, irritation and exhaustion drained away' (see Further Reading). Indeed, in her column in *The Times* newspaper she recommends the report *A Plant a Day Keeps the Doctor Away* produced by the Horticultural Trades Association in conjunction with Reading University and the Royal Botanic Gardens Kew in the UK. So wildlife conservation is not just a sentimental occupation for softies; it's a social duty in which everyone should be involved.

How do gardens differ from the rest of the countryside?

The short answer is that, in principle, they don't – much. This may sound surprising, but most of our planet has been manipulated in some way by humans.

There is very little that has not been brought under control or partially modified by activities akin to gardening. This particularly applies to the UK and, to a large extent, to the rest of Northern Europe.

Gardening is the practice of controlling and coaxing nature into patterns that suit us for our pleasure and enjoyment, or for growing our own food. Human nature being what it is, it's also often for keeping up with fashion, competing, or enhancing social status. For many millennia the same has been happening in the environment at large. Agriculture – be it arable or pastoral – has altered and managed the countryside, whilst owners of large estates have tamed their land and moulded it to suit their tastes. There is very little left which could truly be called 'wild' – perhaps just such habitats as mountain tops, salt marshes, shifting shingle spits and sand dunes. If Europe had been left entirely to itself after the retreat of the last ice age, it would now be mostly forested as the result of **succession** (though recent research suggests that grazing animals might have created a mixed landscape of woodland, scrub and open grassland). 'Succession' is an ecological series of events with which we, as gardeners, are all too familiar. It's what happens if we go away for a couple of seasons, leaving the garden with no one to tend it!

Ecological succession

Nature is constantly on the move. In the words of the popular hymn – 'change and decay in all around we see…'. Consider a pile of bricks left outside in a damp, shady place. Before long they will be colonized by mosses or lichens. Leave them for several years and they will have cracked and crumbled and, perhaps, have grass growing out of the crumbly bits. Leave your gravel path untended for a season and it will be covered with grass and other small plants. Leave it for several years and it will turn into a hedge and, in half a century, into linear woodland! This process has been much studied by ecologists who have coined the term succession

The rest of this chapter relates mainly to the UK, but the *principles* are the same the world over – it's just that the species involved will differ according to climate and geographical location.

Natural succession

Provided that there are no factors holding succession back, there can be a characteristic progression from bare rock to woodland. First an area of smooth, bare rock may become weathered by water, wind or frost to form irregularities or small cracks. In these places enough moisture accumulates to support life, and the first organisms to gain a foothold are lichens or mosses. Lichens are a remarkable association between a fungus and an alga. Algae are primitive, often single-celled, plants, and the cells of the alga lie buried amongst the thread-like cells of the fungus to form a conglomerate, which is the lichen. This is a very convenient association, because the alga makes food from water and carbon dioxide by photosynthesis (see Chapter 1), supplemented by mineral elements from the rock, while the fungus feeds on the organic matter in dead algal cells. Together, in amazing partnership (symbiosis), they are adapted to living in the harshest conditions. Like all living organisms they produce waste products, amongst which are acids which break down the surface of the rock a little further. Mosses, too, may be able to gain a foothold in cracks or rough areas where there is sufficient water. After many years the crumbling away of the rock, the accumulation of dead lichen and moss, the action of bacteria and, perhaps, even mineral dust blowing in from elsewhere, is enough to have made a thin layer of soil.

The accumulation of soil means that seeds can settle and germinate. They may be blown in from grasses and other plants, carried on the feet or in the intestines of birds, or stuck onto the fur of mammals. The first to become established in a layer of soil are usually herbaceous species and grasses; so a flowery, grassy sward develops.

But things don't stay like this for long. Slower-growing shrubby species may have germinated meanwhile. As these grow taller they begin to cast shade over the herbaceous species. There is now fierce competition for light and resources between woody plants and the lower-growing herbaceous ones. The woody ones are more successful and, in ecological parlance, become **dominant**, resulting in a scrubby sort of community – heathland or hawthorn and blackthorn scrub.

Finally, taller-growing trees become established and you have a fully-fledged woodland. Ecologists call this the **climax** because it is a final stage which will remain in a steady state, as forest, unless forces intervene to determine otherwise.

The process of succession can be summarized:

Bare rock > lichens and mosses > grassland > scrub > woodland.

As so often in the living world, there are exceptions. On stony screes you will sometimes see birch trees becoming established as the pioneers, and birch or willow can often be spotted growing out of an old roof or wall. You only have to look closely at railway sidings and walls on the approaches to big city stations and you will see that sturdy shrub, *Buddleja,* popping up in places where there is no soil at all.

The actual species which become established in succession depends very much on the kind of rock on which the process started. The weathered rock supplies the mineral elements and determines the nature of the soil, be it rich in minerals or poor, alkaline or acid. On acid soils in the UK and Northern Europe the characteristic scrubby vegetation would be heathers (*Calluna vulgaris* and *Erica* spp.), broom (*Cytisus scoparius*) and gorse (*Ulex* spp.), producing a heathland or heather moorland community. The pioneering trees would tend to be birch (*Betula* spp.) and rowan (*Sorbus aucuparia*), and the final climax, oak (*Quercus* spp.). On chalky or limey soils the vegetation would be quite different – hawthorn (*Crataegus monogyna*) is likely to be the successful shrub, and ash (*Fraxinus excelsior*) or beech (*Fagus sylvatica*) the trees. Different plant species are adapted to different conditions (see Chapter 4), so it's no good trying to grow heathers in a limey soil. If your garden is on a sandy, free-draining, acid substrate and you abandon it for a few seasons you will come back and find it has turned into a heath – lizards and adders might even have moved in too!

Succession may, of course, start in water; a series of changes ultimately ends with the same kind of climax as for dry land (woodland), though via a different route. In deep water, up to depths through which light can penetrate, submerged plants become established. Towards the shore, as the water becomes shallower, floating-leaved plants such as water lilies can live. Then come the plants which grow round the edges, the emergents, with only their 'feet' in water, such as water plantain (*Alisma plantago*-aquatica), marsh marigold (*Caltha palustris*), reeds (*Phragmites communis*), rushes (*Juncus* spp.) and sedges. As these shallow-water plants shed their dead remains, material accumulates on the bottom and the water becomes even shallower, to the point where it may be dry enough for shrubby species, such as willows, to survive. They are happy to stand in water or wet ground, and may be succeeded by alder trees, which are successful, too, in waterlogged situations. Thus, round the edges, emerge the beginnings of woodland. Meanwhile, the plants of the reedy fringe find that they can move out into water of the right depth for them, pushing the floating-leaved lilies and pond weeds further into the middle.

As this process progresses a body of still water, such as a pond or small lake, eventually turns into dry land. If you're the owner of a garden pond you know only too well that, in order to hold back the succession, you have to periodically do some clearing out. Looking after a pond is, in some ways, *harder* than looking after a flower bed!

Succession in water is given the ecological term the **hydrosere**, and can be summarized by Figure 6-1.

This account of natural succession is a simplified story and there are all sorts of variations on the theme. Types of succession depend not only on the nature of

Figure 6-1. The hydrosere – the successional changes which take place in a body of still water.
(a) Trees and shrubs growing in damp soil.
(b) Reeds growing in shallow water.
(c) Floating-leaved vegetation, for example, water lilies.
(d) Submerged vegetation growing in deeper water.
As time goes on, dead organic matter from (b) settles on the bottom and eventually dries this edge out so that the reeds have to move further out into water – which has meanwhile become shallower due to the accumulation of dead leaves and stems of (c). The (c) plants are now crowded out and have to move out towards the middle. Unless halted by mechanical disturbance this process continues until the pool is dry land and covered with trees.

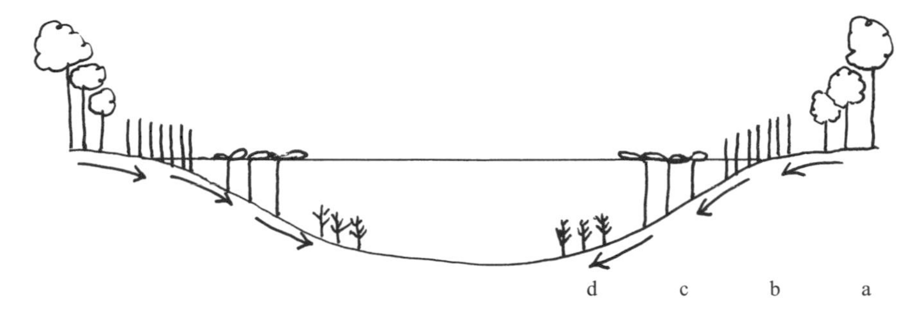

the starting point but also on climate. The kind of forest in northern latitudes is clearly different from that in the tropics. I've already mentioned that trees are occasionally pioneers rather than herbaceous species. The climax may be different, too, when circumstances prevent the succession from taking its normal course.

Some examples of succession held back

The commonest cases of this occur where humans have intervened. Thus, a meadow or pasture is the grassland stage, held there by the action of mowing or grazing; the heathlands of Thomas Hardy's 19th-century Wessex in southern England were the result of light grazing and the cutting of 'furze' for animal bedding and kindling wood. In other parts of the world natural grasslands or savannah result from the grazing pressure of herds of herbivorous animals. In North America the prairies were kept in their steady state by herds of buffalo, and in Africa the savannahs are maintained by such herbivores as wildebeest, antelope and elephant. Even in the natural climax of forest the equilibrium will sometimes be disturbed, for example when a giant tree dies and falls, creating an open glade, which may then be grazed by deer. There may have been a fire, a volcanic eruption, erosion by wind or water as on a coast or the banks of a big river, or slippage on a steep slope. In such situations it may be 'back to square one', and the succession starts again.

Present-day habitats – the results of succession held back

As a result of the process of succession that would have occurred as the last ice retreated from Europe about 10,000 years ago, much of the land would have become covered with forest. The only open, grassy and flowery habitats would probably have been on the tops of the highest mountains, glades in forests where very old trees had fallen, and places along river banks and coasts where erosion and wind had taken their toll. Gradually the forest has been cleared, crops grown, herds of animals put to graze. Some hedges are thought to represent remaining bits of the old wild wood, but over the centuries they have been laid and trimmed, and new hedges planted. When you are enjoying a day out in one of the UK's national parks, such as the English Lake District, or some other attractive environment, you may feel that you're out in the wild, relishing a taste of 'wilderness'. Nothing could be further from the truth. The landscapes of Cumbria, the Scottish Highlands, Snowdonia, the Norfolk Broads and so on are *artificial*, the result of centuries of human activity akin to gardening – the uplands, for example, by forest clearance and grazing, the Broads by medieval peat cutting.

Until World War II this 'gardening' – in other words, mainly farming – was, fortunately, fairly gentle. It actually created *more* diversity than there would have been had the countryside been left almost completely forested. The fields created by the earliest farmers provided habitats where a large variety of different grasses and wildflowers could flourish. Because the soil was not heavily treated with man-made nitrogen fertilizers, the grasses and cereal crops didn't produce very luxuriant foliage and this allowed the more delicate flowering species to compete favourably. In pastures and hay meadows there would have been a profusion of perennial wildflowers flourishing, year after year, along with the grasses – in pastures, for example, cowslip (*Primula veris*), clover (*Trifolium* spp.), and the dandelion tribe; in hay meadows ox-eye daisy (*Leucanthemum vulgare*), knapweed (*Centaurea nigra*) and yellow rattle (*Rhinanthus minor*). In the arable fields there would have been colourful annuals such as corn cockle (*Agrostemma githago*) and poppies (*Papaver* spp.), propagated each year by seed. In many countries of Eastern Europe, where many farms are small-scale, and unable to afford modern methods, beautiful flowery fields (which in the UK would be protected habitats) can still be found in profusion. Ominously, change is in the wind. With entry into the European Community there is going to be more money for modern intensive farming and the old-fashioned fields are in danger of being lost.

After World War II the UK government was keen to make sure that we could be self-sufficient in food – things so nearly came adrift when our imports were threatened by German attacks at sea. A booming agro-chemical industry developed; a great deal of effort was put into breeding the leafiest of grasses, the most productive of vegetables and cereals, and farmers were subsidized to modernize their methods. The results were the ploughing up of old fields, the sowing of monoculture grass leys which were heavily fertilized, drainage of wetlands and the elimination of 'weeds' on arable land. More recently economic pressures have forced farmers to cut back on labour and produce as much as possible from their land by mechanical and artificial chemical means. Hedges have been removed to create enormous fields able to accommodate the manoeuvres of giant machinery, and the use of profligate amounts of artificial fertilizer and pesticide has continued. In short, diversity in the countryside has been declining dramatically. Travelling by rail or road from London to Manchester in spring, you look out on a green and pleasant land – but, in ecological terms, it's a desert. Those bright green fields support only swathes of boring rye grass or vast hectares of winter wheat – the price we've paid for a plentiful supply of cheap food.

During these decades of agricultural progress, nature lovers and the general public became ever more alarmed at the drastic losses of wild species. Enter then the wildlife conservation organizations. In the voluntary sector, the Trusts for Nature Conservation (now the county Wildlife Trusts) were founded, the Royal Society for the Protection of Birds became more active, Friends of the Earth campaigned, and so on. In the statutory sector, the Nature Conservancy Council (subsequently English Nature, and now being incorporated into a wider conglomerate of countryside management agencies), was responsible for improved legislation to protect valuable sites. From the 1970s onwards stalwart efforts have been made to turn back the tide of increasing destruction, by purchasing land to create nature reserves, by legislating to protect the most precious areas and by trying to encourage farmers to be 'nature friendly'. Recently the National Trust, in conjunction with United Utilities and the Forestry Commission, has set up a project to return Ennerdale, in the English Lake District, to its original wild state – an undertaking which may take several hundred years. But it has been a struggle – some would say even a losing battle.

This, though, is where domestic gardens can make valuable contributions. To recap, our gardens are not very different from the countryside at large, in terms of 'managed environments'. So it makes sense to treat them as potential nature reserves. I've already mentioned scientific work which has demonstrated how rich in wild species a conventional garden can be. What follow are some ideas about replicating in gardens some of the features of the diverse countryside of old-fashioned pre-war farming.

Types of habitat which can be created in gardens

There is plenty of excellent literature which provides detailed guides to so-called 'wildlife gardening', but it might be useful to summarize the kinds of things you can do to create the disappearing features of our formerly rich and varied landscapes.

Ecologists love classifying things and have spent a great deal of effort trying to impose order on our many different kinds of plant and animal communities. There is now a monumental work, the National Vegetation Classification, which categorizes all existing vegetation types in the UK. It goes into enormous detail, and some would argue that nature cannot be so rigidly regimented. Be that as it may, some broad categories can be easily recognized and are summarized below. I emphasize that this is a brief overview of the habitats it is possible to create in a garden, and some suggestions for action; for more information see Further Reading.

A warning shot across the bows for would-be 'wildlife gardeners'

Be warned! Wildlife gardening is just as hard work as is any other kind of horticulture. Land management often creates diverse habitats, so what you might be doing in your wildlife garden is the equivalent of farming or forestry! Rudyard Kipling's verse is just as applicable to the business of maintaining semi-natural ecosystems in the garden as it is to conventional gardening.

Our England is a garden, and such gardens are not made
By singing: 'Oh, how beautiful!' and sitting in the shade,
While better men than we go out and start their working lives
At grubbing weeds from gravel-paths with broken dinner-knives...

(Reprinted by permission of A.P. Watt Ltd. on behalf of The National Trust for Places of Historic Interest or Natural Beauty)

Kipling's sentiments are particularly pertinent for me because I once found myself having to resort to a knife to control a thick mat of rhizomes, which a native species of pond weed had formed on the bottom of my pond. It was seriously hazardous for the welfare of the pond liner, but eventually successful after a long, hard struggle.

But don't be put off – you'll be rewarded by the pleasure of sharing your garden with creatures you might not otherwise have the opportunity of seeing, and that warm 'doing good' feeling of helping to conserve our planet's biodiversity!

Unfortunately, some of the rarest and most threatened habitats, for example peat bog and heathland, don't lend themselves to re-creation in gardens. There are, however, three broad categories of habitat (with subdivisions) which it is feasible to create in an average-sized patch:

- woodland
- grassland
- small-scale wetland.

Woodland

In the UK there is no truly wild, original forest; it has all been modified by human activity, be it for timber for the navy, to provide coppiced wood for charcoal-making, or for grazing as wood pasture. So, botanists and foresters call woodlands which have not been artificially planted **semi-natural**. They fall into two categories – **primary woods** on sites which, as far as we know, have always been wooded, and **secondary woods** where trees have naturally colonized, or have been planted, on bare ground (perhaps formerly agricultural) in historical times. These woods are mainly composed of native species, those which have become established without human aid in the British Isles during the past 10,000 years or so. This kind of woodland supports the richest variety of other plants and of associated animals. **Commercial plantations** are a third category of woodland. These are stands, very often of non-native species such as Sitka spruce (*Picea sitchensis*) or Douglas fir (*Pseudotsuga menziesii*), grown for commercial forestry purposes. If you see trees closely planted, maybe in straight rows and all looking much the same age (same height and girth), you can be pretty sure it is such a plantation.

The aim of creating a new woodland for wildlife should be to get as near as possible to semi-natural woodland. If you are fortunate to own a large patch of land then it may be possible to plant a small woodland but, in the average garden, it's not practicable. What *is* possible, though, is the creation of **woodland edge** habitat.

Structure of a woodland

Ecologists are great classifiers and like to describe a woodland in terms of various layers, a bit like the floors of a building. The tallest trees form the top stratum – the **mainstorey** – the top branches of this spreading out to form the **canopy**. Below this there may be an **understorey** of smaller trees, and then the **shrub layer** of bushy species between 1 m and 2 m tall. Under this is the **herb layer** which supports grasses and herbaceous plants, and the **ground layer** is bare or covered with mosses and liverworts. Within these layers there are sometimes recognized further sub-divisions – for example the dwarf shrub layer between the shrubs and herbs.

Figure 6-2. Woodland structure.

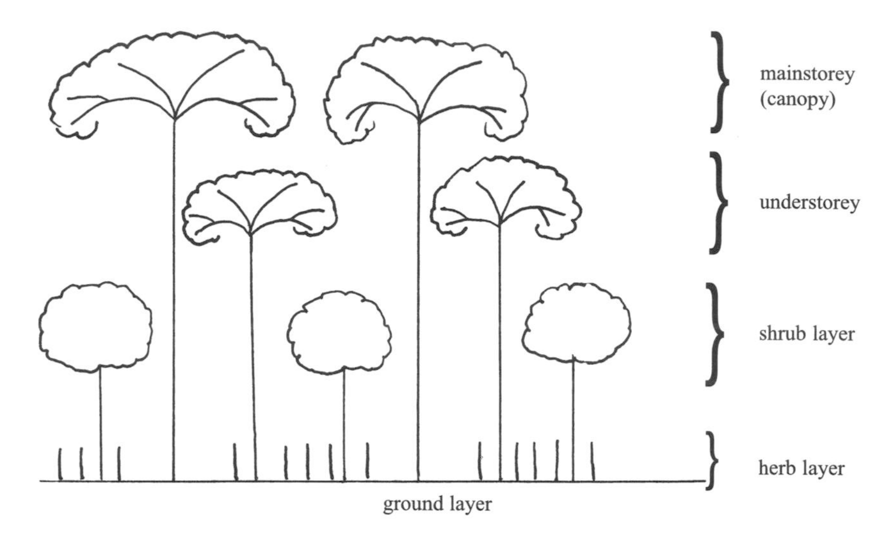

Not all forests necessarily contain all layers. In a tropical forest they could probably all be recognized, but in a planted British beech wood, for example, there might be only the mainstorey and a ground layer. Beech trees have a very closely structured canopy (look at how the fairly thick, tough leaves are arranged), allowing very little light through to the ground, and discouraging the growth of shrubs and herbaceous plants.

At the edge of a wood – particularly if it is south facing – there is more light available and more opportunity for herbaceous plants on the ground, and an intermediate layer of shrubs, to flourish. All the layers are present in profusion. A hedge is really a double woodland edge. So in a garden, by creating a sort of hedge structure, if possible with some taller trees in the background, you are replicating a woodland edge habitat.

The kinds of plants found in a British woodland edge are native shrubs such as elder (*Sambucus nigra*), dogwood (*Cornus sanguinea*), guelder rose (*Viburnum opulus*) and bramble (*Rubus fruticosus*), and flowering herbaceous species such as primrose (*Primula vulgaris*), bluebell (*Hyacinthoides non-scripta* – the narrow-leaved native variety, not the more aggressive and invasive continental version) and foxglove (*Digitalis purpurea*).

Gardening suggestions

Trees add not only structure and interest to a garden but also, from the point of view of encouraging wildlife, provide food and nesting sites. So, if at all possible, have some trees as the background to your woodland edge habitat. The recent Sheffield University research, the BUGS project (see Further Reading), has shown that the presence of trees in a garden is a major factor in increasing biodiversity. If your garden is small, say less than 0.25 ha (about half an acre), then large forest trees are not a good idea. (There are some suggestions for small trees in Appendix 1.)

The usual advice is to stick to native species and to aim at semi-natural woodland. This is because native trees and shrubs tend to have a good variety of native insects associated with them. Plants from other countries have been imported minus their native fauna and, in any case, many of the invertebrate species from other climates might not have coped with the new conditions. Studies on the insects associated with various tree species have shown that the native oaks (*Quercus robur* and *Quercus petraea*) come out top, with over 300 species depending on them for food or breeding sites. Unless you have a very large garden, oak is not really practicable (though it will coppice – see page 112). Native willow species (*Salix* spp.) are almost as good (see gardening suggestions below), and hazel (*Corylus avellana*), birch (*Betula* spp.) and alder (Alnus *glutinosa*) are useful, both for pollen in spring and their nuts (hazel) or seeds (birch and alder) in autumn. But, contrary to some of the purist literature, there is no need to adhere religiously to native species (see page 110). Non-native conifers such as *Cupressus* spp. provide good cover for birds in winter and nesting sites in spring. But beware of the fast-growing *Leylandii* beloved of surburban hedge owners. If you have inherited *Leylandii*, which for some reason cannot be removed (as have I, in the form a long border hedge) don't despair. You'll have to keep it under control but it isn't necessarily a disaster – dunnocks, blackbirds and house sparrows may nest in it.

Native trees and shrubs providing berries and seed

Species with berries – holly (*Ilex aquifolium*), rowan *(Sorbus aucuparia*), hawthorn *(Crataegus monogyna)*, blackthorn *(Prunus spinosa*), elder *(Sambucus nigra*) and guelder rose *(Viburnum opulus*) are great as a food source for blackbirds and winter-visiting fieldfares and redwings. I've observed that guelder rose berries, bright red and wonderfully shiny as they are, are not favourites and are often left until the very last. But one day I looked out of the window onto our small native-species plantation in Yorkshire to behold a large flock of waxwings descending on the remaining guelder rose berries. They were devoured in an instant!

Birch (*Betula pendula* or *pubescens*) and alder (*Alnus glutinosa*) provide pollen in spring and seed in autumn, while hazel (*Corylus avellana*) is a good pollen source and provides nuts in autumn. Willows (*Salix caprea* – goat willow or 'pussy willow') and (*Salix cinerea* – grey willow) are an excellent source of pollen early in the year; but you must have the male bushes – trees of the willow family are dioecious (male and female flowers on different plants). Ivy (*Hedera helix*) flowers late in the year and provides supplies of pollen in early to late autumn. There is no need to worry about ivy growing up shrubs and trees, as long as it is not smothering their leaves. Foresters only dislike it because of its tendency to make branches top heavy and increase the danger of wind damage.

Coppicing trees

Some forest trees are quite masochistic and very amenable to being strictly controlled. You can be as severe as you like, cutting them down to a foot or so above the ground, and they will sprout again quite happily. (Anyone who has tried to get rid of a sycamore will be familiar with this phenomenon!) Coppicing was a common practice in pre-20th-century Britain – a source of fencing posts, brush handles, small logs for charcoal making, and so on. See Appendix 1, page 152, for species that can be treated like this and maintained as small trees or shrubs.

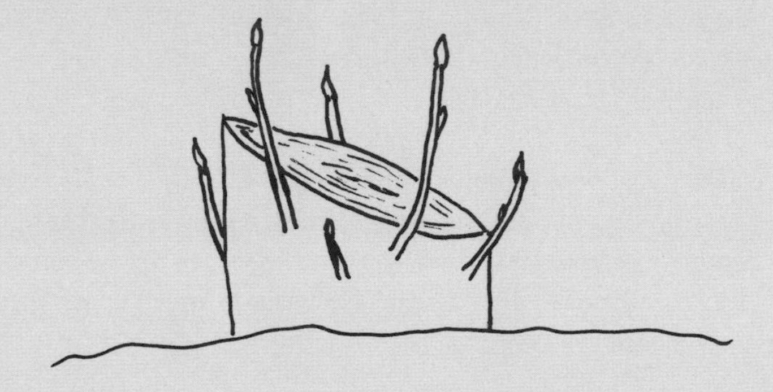

Figure 6-3. Coppicing. A tree trunk has been cut close to the ground (the sloping cut helps to stop water accumulating in the stump and protects against decay). Shoots soon emerge either from the visible stump, or from just below the ground. The stump was traditionally called the 'stool'.

Shrubs

If you are fortunate enough to have room for trees it's a good idea to try to replicate the shrub layer below them. If your garden is only going to allow a hedge or a shrubbery then shrubs are your starting point. Appendix 1 also contains a list of suitable native shrub species.

The herb or 'ground' layer

Here there will be grasses and herbaceous species which can tolerate shade or partial shade. If you choose carefully this can be colourful throughout the year – for example, daffodils (*Narcissus* spp.) snowdrops (*Galanthus nivalis*) and bluebells (*Hyacinthoides non-scipta*) in spring, and foxgloves (*Digitalis purpurea*), red campion (*Silene dioica*) and herb robert (*Geranium robertianum*) later in the summer. See the lists in Appendix 2.

Aim at a layered effect

As you look at your woodland edge habitat you shouldn't be able to see any bare ground, but the three layers should be visible in sequence – herbaceous species at the base, then shrubs and then, if you have space, trees.

The ongoing management

Things don't stay the same. In your woodland edge the trees and shrubs will grow bigger, and the herb layer will tend to be invaded by tree and shrub seedlings. You'll probably have to work fairly hard to top and trim the trees, certainly to keep the shrubs under control and to weed the herb layer. This is particularly so if your garden is small.

Letting dead wood lie

In a natural woodland branches fall off trees and lie on the ground. They gradually become colonized by all manner of micro-organisms and invertebrate creatures: beetles which lay their eggs under the bark where the grubs hatch and feed, fine threads

The woodland edge.

(hyphae) of fungi which penetrate the wood and break it down, providing bits and pieces which are consumed by woodlice and millipedes, spiders preying on smaller creatures, to name but a few. This provides a cornucopia of food for birds and small mammals such as hedgehogs and shrews which, if encouraged, can be useful to you in feeding on garden pests (see Chapter 5). No one applies fertilizer in a natural forest and yet the trees grow and flourish; so let dead wood lie and even create log piles. These, as well as providing food organisms, might have the additional advantage of providing a hibernating place for hedgehogs.

Laying a hedge

In the days when there was plenty of cheap farm labour, farmers would manage their hedges, not by chopping them back haphazardly but by skilfully nicking the largest trunks low down, laying them horizontally and weaving other branches into them. Woody plants, resilient as they are, sprout again quite happily and this structure gives good thick growth at the base. So if you have a hedge which has to function as a boundary, why not learn to lay it? In the UK the National Hedgelaying Society and the British Trust for Conservation Volunteers run demonstrations and courses.

Dead leaves

Ecologists call them collectively **leaf litter** – what you might call 'desirable rubbish'. I have often wondered whether tidy-minded aliens from outer space would regard autumn leaves on the ground in the same light as crisp packets and coke cans: beauty, and therefore presumably ugliness, is in the mind of the beholder. We somehow instinctively don't mind the odd leaf around – though their banana-skin effect on roads, paths and pavements is a hazard to be eliminated. If you're a neat and tidy gardener you'll want to remove them from lawns and paths but, elsewhere, they are really rather desirable. I'll forgive you if you clear them from paths and patios, mouldy ones from rose beds and under diseased fruit trees, or conifer needles and *Leylandii* clippings

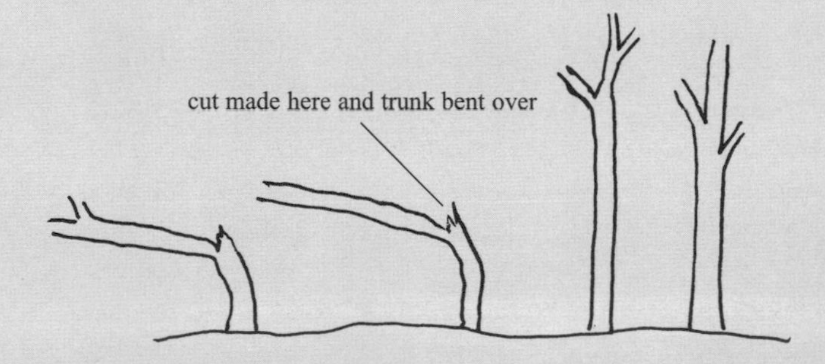

Figure 6-4. The principle of hedgelaying. The treatment encourages the trunk to sprout from the base and so gives a nice thick hedge from the ground upwards.

which don't decompose easily. But leaves from healthy broad-leaved trees should definitely be left to lie in your shrubbery or woodland edge wildlife habitat. There is a hive of activity in the leaf litter on the woodland floor which not only supports some of the above-ground food web but also eventually returns nutrients to the soil via the **decomposer food web** (see earthworms and organic fertilizers in Chapter 4). As well as enriching biodiversity all this activity on the woodland floor recycles mineral nutrients for use by the trees and shrubs - so if you have a rich layer of dead leaves there's no need to fertilize.

Grassland

Grassland tends to fall into two types, depending on underlying conditions, and both supporting different plant species: hay meadows and permanent pasture.

Hay meadows

In the past, farmers would have left certain fields to grow through the spring and summer. These meadows were often in damper low-lying areas and contained a type of grassland which supported a variety of grass species intermingled with many other herbaceous perennials – mostly those which grow quite tall and flower in late spring and early summer.

The practice of making genuine hay has all but died out, and what you see in fields nowadays are piles of gigantic black plastic bags containing wet grass. This has often been cut much earlier in the summer and mixed with substances, including molasses – responsible for the sweetish sickly smell – which help to preserve it. This is silage which, in effect, is pickled grass. There are two unfortunate things about silage from the point of view of wildlife conservation. One is that hay meadows have been converted into monocultures of one or two species of fast-growing grass with luxuriant foliage. The other is that the grass is cut early in the summer when the leaves are fresh and juicy. This is when ground-nesting birds such as skylark, plover and partridge, and mammals such as the brown hare, are rearing their families. Enough said about the ensuing slaughter. Not only has diversity of plants been lost but the practice is also a threat to wild animals.

Making a flowery hay meadow is fashionable in 'natural gardening' circles but it is much easier said than done. There is also a lot of confusion between a true hay meadow which is based on perennial plants and the kind of flowery field which contains mainly annuals and was originally the kind of weedy, arable land which today one never sees on a modern farm. The Gardening suggestions overleaf should give you some guidelines and clarify the difference between a meadow and a collection of arable weeds.

Gardening suggestions

Choice of plants

Packets of wildflower seeds are now routinely on sale in garden centres and nurseries and there are specialist firms which supply wildflowers, as seed, plugs or mature plants. It's important to check out whether they are meadow plants suitable for establishing in a grass sward, or whether they are annuals, better suited to disturbed ground. For your meadow you need perennials adapted to an open habitat. See Appendix 3 for a list of suitable species.

Techniques for creating a successful flowery meadow

There are two golden rules:

1. The less fertile the soil the better. It is not much good trying to establish wildflowers in a piece of lawn that you have previously treated with nitrogen fertilizer. This is related to the disappearance of flowery meadows in the countryside – the application of nitrogen fertilizers encourages luxuriant leafy growth of grasses and the more delicate herbaceous flowering species lose out in competition with them. It may be necessary, therefore, to strip off turf and some of the topsoil before starting. Alternatively, see if you can find an unproductive corner of your garden.
2. Be prepared for a different cutting regime from the one you are used to. You've got to learn to be a haymaker! Allow your meadow to grow, flower, and set seed before cutting. But you can't just leave it to die down – it must be cut and the 'hay' removed. If not the decomposing remains will enrich the soil again and encourage the invasive grasses. The mass of dead vegetation will also prevent the less vigorous species from getting going in the spring. The cutting may present a challenge, particularly if you only have a small hover-type mower. A strimmer is probably the answer for the size of area you are likely to have in an average garden. Alternatively why not learn to hand-scythe the traditional way? It's excellent exercise (and tremendously good for the waistline)!

A flowery meadow can be created from scratch by sowing seed on prepared ground. It must be fairly infertile and free of perennial weeds, such as dock and thistle. It is also possible to introduce pot-grown plants into turf, or to create bare patches onto which seed can be sown. Both methods require careful management in the first year – regular light mowing to make sure that grasses don't dominate. You probably won't see much flowering in the first season. There is also a decision to be made as to whether you want an early-flowering meadow or a later-flowering one with a different mix of species. The species you choose will also depend to a certain extent on your type of soil. See Further Reading for more assistance.

Creating a wildflower meadow is not simple, and requires hard work and perseverance. But don't be put off! It is *so* rewarding, and is of enormous value to wildlife conservation. You'll be able to sit out there in summer and listen to the humming of bees and maybe the scratching and clicking of grasshoppers or crickets – a sound all too rare in the countryside now. Many other invertebrate creatures, too, will come to live in your hay meadow. Some of them may be attractive – if you have an early-flowering meadow, supporting the pretty lady's smock (*Cardamine pratensis*), you might be blessed with orange tip butterflies which lay their eggs on that plant. There are butterflies which need meadow grasses on which to lay their eggs, such as the meadow brown. You might also be encouraging useful carnivorous insects and spiders which will prey on pests. For me, nothing is more attractive than an area of longer grass, dotted with colourful flowers, and with mown paths weaving through it, leading to and from a traditional close-mown lawn.

Permanent pasture

This is the kind of plant community which can still be found in some upland regions, or in marshy areas not suitable for hay-making. Before the days of artificial fertilizers pastures would have contained species other than grasses. Many would have been the low-growing kind with rosettes of leaves at their bases, for example, daisy (*Bellis perennis*), dandelion (*Taraxacum* spp.) and

The wildflower meadow.

plantain (*Plantago* spp.), or ones with runners which spread out over the ground, like slender speedwell (*Veronica* filiformis), clover (*Trifolium* spp.) and self heal (*Prunella vulgaris*). These plants are well adapted to being constantly nibbled because their growing points are low down. In order to maintain these fields in good condition the grazing would have been carefully controlled so that the pressure on the plants was not so heavy as to damage their capacity to go on growing year after year. Nowadays many of these pastures have either been ploughed and re-seeded, or enriched with nitrogen fertilizers so that grasses have out-competed the less leafy flowering species. In upland areas in the UK, where farmers have had subsidies per head of stock, *over*grazing has been rife – visit the sheep walks of the Lake District, the Yorkshire Dales or the Welsh uplands and you are actually in the midst of a degraded landscape – an ecological slum.

In a garden, the equivalent of a pasture could be a patch of grass with flowers in it which, at periods during the summer, is permitted to escape the lawn mower and grow a little longer to allow flowering. My own lawn has a lot of self heal and clover which I allow to flower in midsummer – and the bees love it! (See Appendix 4.)

Collections of arable annual 'weeds'

This kind of plant community is quite different from a genuine hay meadow. The flowers common in the cereal fields of medieval times were such species as poppies, corn cockles and corn marigolds. They are wonderfully jolly, colourful and an asset to any garden.

Gardening suggestions

Ease up on the mowing

If you have inherited a garden, about which you know relatively little, your lawn may have species in it which will give you pleasure if you allow them to flower. Allow the grass to grow a bit longer at intervals during the summer and see what appears. If there are interesting things – as I have discovered in my own patch of grass – let them show their faces.

Sowing wildflower seed

If you buy a packet of wildflower seed, check that it contains only 'cornfield weed' species. These annuals should perhaps be kept to a flower bed or to your vegetable patch, not sown in amongst grass.

Wetlands

During the fifties and sixties farmers in the UK were not only being encouraged to modernize in terms of applying artificial fertilizers and pesticides, but were also being urged to drain wet areas to bring more land under cultivation. Marshy fields began to disappear, and with them the particular kinds of plants and animals adapted to wetlands. Before World War II there would have been places in river valleys where the watersides were ablaze with bright yellow marsh marigolds (*Caltha palustris*) and the fields purple with fritillaries (*Fritillaria meleagris*). Villagers would have been able to gather armfuls of them. Though they have been introduced successfully into damp clay soils, there are now only a dozen or so places in the UK where wild fritillaries grow in their original natural habitats in any profusion.

Another feature of farmland to suffer was the field pond. Pre-war, several would have been found on every farm. With better piped water supplies to fields, ponds became redundant; so the favoured habitat of the common frog has gradually shrunk. Frogs like to lay their spawn in shallow, often seasonal, pools where there are few, if any, fish. Deeper pools, lakes and gravel pits (one of the post-war gains in terms of wetlands) are not much good for them. Gardeners can make a valuable contribution to reversing this decline by making ponds and maintaining wet patches.

Ponds

There are three important characteristics of a pond rich in wildlife:

- A varied profile at the edge, with at least one place where there is a 'beach' – a gradually sloping shore where frogs, newts and toads can crawl in and out, and where birds can drink and clean themselves. Shallows are also important for establishing marginal vegetation, much of which is very colourful and attractive.
- A place in the middle at least 50 cm deep where water is unlikely to freeze; somewhere water creatures can hide away and perhaps bury themselves in the mud in the hardest of winters. It is best, if possible, to make the central depth greater than 50 cm (but that is difficult to combine with shallower shores in a small pond).
- Plenty of submerged plants which provide habitat for a myriad of invertebrate creatures and also help to oxygenate the water during daylight hours.

Pond problems – real or imaginary

Ponds vary in their characteristics, in the kinds of plants and animals they support, and problems to which they are prone. This is something not always

Figure 6-5. A pond profile, showing examples of suitable plants for a garden pond: (a) water milfoil (*Myriophyllum spicatum*); (b) water starwort (*Callitriche* spp.); (c) amphibious bistort (*Persicaria amphibia*); (d) water crowfoot (*Ranunculus aquaticus*); (e) arrow head (*Sagittaria sagittifolia*); (f) marsh marigold (*Caltha palustris*); (g) flag iris (*Iris pseudacoris*); (h) purple loosestrife (*Lythrum salicaria*).

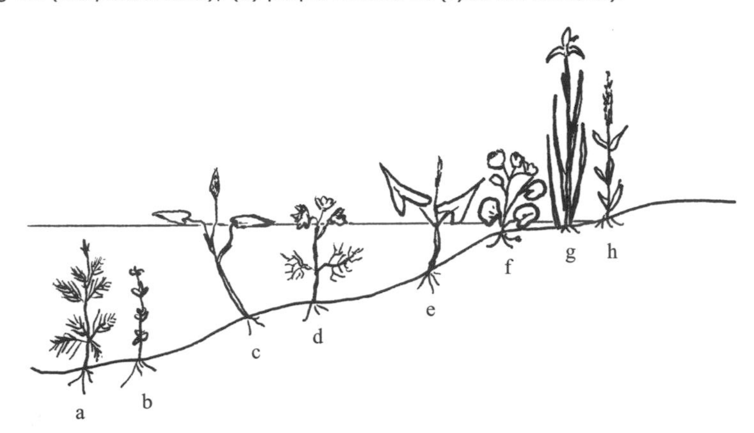

appreciated; wildlife gardening literature often treats ponds as though they were all the same, and some of the potential problems described are not as serious as they would appear.

One important factor in determining the characteristics of a pond is its aspect. In full sun a pool has the potential to support a wide variety of plants and animals, but there are potential hazards to do with overheating. On a hot summer's day the water may heat up considerably, and it is a fact of physics that hot water holds less oxygen than cold (think of the bubbles that escape as you heat water in a pan or kettle). This presents no problem during the day when the submerged plants will be producing oxygen, which dissolves in the water, as a by-product of photosynthesis. But at night not only will all the invertebrate animals, amphibians and fish be consuming oxygen to keep their respiration going, the plants will be taking it up too (see Chapter 1). In addition the millions of bacteria feeding on dead remains on the bottom will be respiring and using up the vital molecules. This represents considerable depletion of oxygen. The gas normally dissolves at the surface and diffuses downwards, but this doesn't always occur fast enough to replenish supplies being consumed below. The result may be that some animals die from 'suffocation', particularly larger, active ones, and those which cannot take in oxygen at the surface. So a little shade, to allow the water to cool down before nightfall, is no bad thing – perhaps situate your pond so that it is shaded in the afternoon.

If the only place where you feel able to make a pond is in a very shady corner under trees *don't despair*! – it will be better to make one there than not at all. It may not support many plants, or as great a variety of animals as one which gets

Figure 6-6. Animals found in shaded ponds containing dead leaves from overhanging trees.
(a) The lesser water boatman, *Hesperocorixa sahlbergi.* This is one of many similar-looking species, but this particular one has very dark markings on its wing covers. It is now relatively rare and is usually only found in dark ponds and reed beds at the edges of lakes.
(b) This caddis larva, *Glyphotaelius pellucidus*, lives inside a case constructed of carefully cut pieces of dead leaf.
(c) The water hog louse or water slater, *Asellus* spp., (related, as its appearance suggests, to the wood louse) is commonly found in places where there is a lot of dead organic matter.

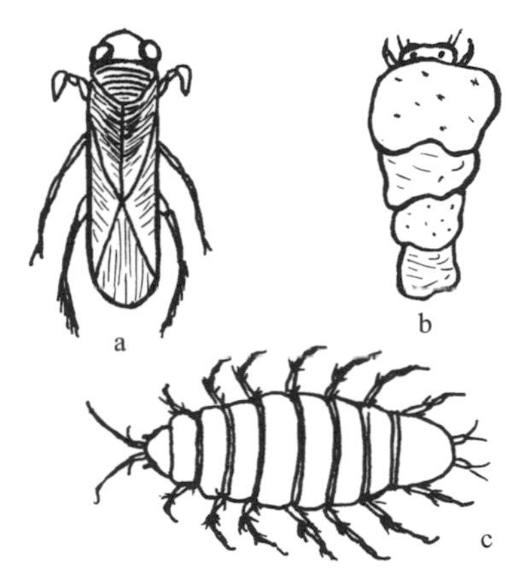

plenty of light, but my research (unpublished) has shown that there are certain species which quite like a dark and dingy habitat with dead leaves on the bottom and the odd dead branch to lay eggs on, or to cling to. Examples of such creatures – not to be altogether despised – are the water slater or water hog louse (*Asellus* spp.), a particular species of lesser water boatman, (*Hesperocorixa sahlbergi*), and even a few species of caddis fly (*Trichoptera*) – particularly the ones whose larvae make their cases out of dead leaves or tiny bits of stalk or twig.

Another important factor is the nutrient content of the water. If it is too rich in nitrates and phosphates the rapid growth of algae will be encouraged and you will end up with a lot of green slime. Nutrient-rich water will also boost the growth of floating plants such as duckweed (*Lemna minor*). These are real problems; see below for suggestions for dealing with them.

Gardening suggestions

Read a good guide to pond-making before starting

There is too much detail to go into here, but there is plenty of literature with good advice (see Further Reading).

Avoid ready-made plastic ponds

These invariably have steep sides and don't allow for shallows and beaches.

If a pond is not possible

Even the smallest water feature is worthwhile; an old sink with a bank of pebbles at one side, or an upturned dustbin lid sunk in the ground and kept full of water, useful as a bird bath. They are both better than nothing.

Are fish desirable?

Wildlife gardening guides often advise that you should on no account introduce fish into your wildlife pond. The reasoning is that they will gobble up any frog, newt or toad spawn and prevent your pond being a haven for amphibians, a group of animals which needs conserving. The fossil record tells us that fish were already well established when amphibians started to evolve – all the evidence suggests that they evolved from a rather 'leggy' kind of fish which could survive for periods out of water. This means that there must have been severe competition between fish and the earliest amphibians. So shallow water, or pools which periodically dry out, where large fish were unlikely to be present in large numbers, might have been advantageous for them (and also easier for them to pursue their habit of emerging from water onto land during their life cycle). On the other hand, frogs lay such large quantities of spawn that there should be sufficient for the odd goldfish dinner with enough left over to grow into tadpoles and adults. In any case, your goldfish are going to have to run the gauntlet of visiting herons (live and let live!). So one or two goldfish aren't going to do much harm but roach, tench, bream or carp are not a good idea – unless you have so much land that you can afford to have both a fish pond and one for amphibians.

What to do about dead leaves

It is commonly advised that ponds should be covered with netting during autumn leaf fall to prevent dead material accumulating and the bacterial activity of decay causing de-oxygenation. This is a good idea if your pond is in a sunny position where the water will get warm and decomposition will be rapid (bacteria work faster the warmer it is). But if it's in a cool, shady place decomposition will be slow and there is no need to worry too much. Clearly one wouldn't want the pond to fill up entirely with dead leaves so occasional partial clearance would be sensible, preferably from the surface as they fall to avoid stirring up mud from the bottom. Leave the rakings on the side of the pond for a few hours to allow any creatures to crawl back into the water.

Duckweed and other floating invaders

The floating duckweed (*Lemna minor*) is a genuine problem because it prevents light from reaching submerged plants. I once thought I had been so careful, having meticulously inspected and washed a water lily plant, brought from a friend's pond where I knew there was duckweed. But I can't have been sufficiently assiduous because, within a few weeks, a mat of green was spreading over the surface of our new pond. This tiny floating plant thrives on phosphate-rich water and, indeed, our pond was sited downhill from fertilized agricultural fields. It reproduces exponentially, doubling its number per unit time.

Another similar plant is the alien water fern (*Azolla filiculoides*). The best solution for both these problems is to keep scooping out, in spring and early summer, with a long-handled net. (One can be made from curtain netting, a coat hanger and a long cane.) If you allow them to take over they will reduce the amount of light reaching the submerged plants and these will suffer accordingly.

Green slime is another nightmare which plagues pond owners, made worse if the water is nutrient-rich and the pond is in full sun. This green stuff is usually a filamentous alga, *Cladophora* or *Spirogyra*. Prevention is better than cure and a preventive measure is to use rain water, virtually free of nutrients, collected in water butts (every garden should have them), for the initial filling of the pond. I know you're going to say it's going to take an awful lot of water butts to do the job, and that's true! A good way of filling your pond is to fix the end of the hose to a piece of wood. This will float on the surface and so prevent the cascade of water from disturbing the bottom, making the water turbid and stirring up nutrients.

If all this has not been possible, and you are plagued with algae, science has provided a solution in the form of **barley straw**. Experiments have been done to investigate the effect it has on the growth of algae. The mechanism whereby it inhibits growth isn't yet understood, but biologists have worked out the optimum quantities to use in order to achieve good control, whilst preventing rotting straw accumulating on the bottom. The straw can be purchased in net bags and an initial application should be 50 g per square metre of surface area. When this has disappeared quantities should be reduced to 25 g and then 10 g per square metre.

You may have read recently that there has been some doubt about the safety of barley straw but these doubts have now apparently been dispelled.

Aggressive pond weeds

Some native pond weeds are very invasive and tough, making their control difficult. Stick to ones that are easy to weed out when it comes to clearing-out time.

Submerged	
	Starwort (*Callitriche* spp.)
	Spiked water milfoil (*Myriophyllum spicatum*) NB Not the alien parrot's feather (*Myriophyllum aquaticum*) which will rapidly take over.
	Rigid hornwort (*Ceratophyllum demersum*)
Floating-leaved	
	Common water crowfoot (*Ranunculus aquatilis*)
	Water soldier (*Stratioides aloides*)
	Water violet (*Hottonia palustris*)

NB If you are keen to have water lilies, or other invasive species, plant them in baskets to make them easier to control.

Pond clearance

As already implied, looking after a pond can be as much hard work as tending a flower bed. When the time comes for clearing out some vegetation (the inevitable result of succession) do it in sections in small doses, so that you don't upset the animal inhabitants too much. Leave the extracted water weed on the side for a few hours to allow any creatures to crawl back into the water – most of them, even the insects, are capable of doing this.

The garden pond.

Marshy areas

In ecology books you will find wet and soggy areas of land referred to variously as swamps, marshes, mires, fens, blanket bogs, raised bogs, carrs – it's all a bit confusing. The terms swamp and marsh refer to areas which are flooded for all or part of the year and which have a substrate of mainly mineral material. There are also various types of wet grassland – water meadows, grazing marshes, and so on. Mires are based on peat, dead plant material which has never decomposed properly. Bogs, fens and carrs are sub-divisions of mires. A carr is a wet area with trees. The difference between a bog and a fen is to do with the pH and mineral content – bogs being acid and fens less so.

In a garden it is unlikely that one would want to be so purist as to create these different types, not least impractical. But because so much wet land has been drained, it is really valuable to create a patch where at least some wetland plants can grow. Examples of beautiful ones are marsh marigold (*Caltha palustris*), purple loosestrife (*Lythrum salicaria*), meadowsweet (*Filipendula ulmaria*) and great willowherb (*Epilobium hirsutum*).

Gardening suggestions

Creating a marshy are

Your pond liner can be extended beyond the edge of the water body proper, to create an area which will collect rain water and won't easily dry out. (If the dry summers we have been having recently in the UK continue, drying up of ponds and wet areas is going to be an increasing problem.) Here you can plant the many attractive wetland species recommended in wildlife pond guides.

Peat conservation

The peat sold as a horticultural product is obtained from peat bogs - a valuable and threatened habitat. Although you might argue that there is enough peat in Ireland to go round for a long time to come, some areas in the UK have been devastated by peat extraction. One example is the vast stretch of land just south of the Humber, Thorne and Hatfield Moors, which at one time was pristine acid bog land, supporting a diverse flora and fauna with many rare species peculiar to that kind of habitat. The peat has been systematically 'mined' by a firm supplying pure peat and peat-containing composts until there is scarcely any natural habitat left. If you stick to peat-free composts, which are now improving in quality, it will help to conserve rare plants such as the sundew (*Drosera* spp.) and butterwort (*Pinguicula* spp.), and beautiful insects such as unusual butterflies and dragonflies.

Encouraging biodiversity in a conventional garden

It is still possible to have a normal-looking garden in which wildlife is welcome and flourishing. By this I mean containing mainly the usual cultivated garden plants. It helps, though, to select suitable plants and to not be paranoid about tidiness. You'll be pleased to hear that the commonly suggested idea of keeping a patch of nettles as a breeding site for certain butterflies has been shown to be ineffective in an initial piece of research in Sheffield gardens (the BUGS Project – see Further Reading).

Trees and shrubs

Research has shown that trees in any kind of garden are of considerable benefit in increasing biodiversity. It is generally considered that native trees and shrubs are best for invertebrates and birds, but they can be mixed with more conventional garden species. Non-native conifers such as *Cupressus* spp., rhododendrons and laurels provide good cover for birds in winter and nesting sites in spring. There are several cultivated non-native shrubs too, such as *Pyracantha, Cotoneaster* and *Berberis*, which supply a plentiful menu of berries. *Mahonia* provides good cover.

Border plants

You can have a very attractive conventional border in which all the plants in some way benefit insects and birds.

Lists of border plants recommended for attracting insects vary widely. It's also worth distinguishing between plants on which larvae feed and plants which supply nectar and/or pollen for adult insects. Adult butterflies, moths and common wasps need nectar, while bees need both pollen and nectar (pollen for feeding grubs and immature adults and nectar for themselves as adults). I mention wasps because they are not all bad! Their larvae are carnivorous and so the adults collect small insects which may be pests.

There are also herbaceous border plants which can provide seeds for birds such as greenfinches, goldfinches, sparrows and siskins.

Gardening suggestions

Useful plants

Be observant and look out for the plants in your garden – or in gardens you visit – which seem to attract insects or birds. Go out at dusk, or in the dark with a torch;

find out where the moths are feeding, and make a note to plant more of whatever it is they like. Observe during the day which plants the bees favour, and in autumn whether any birds root around for fallen seeds.

See Appendix 7 for a brief list of the ones I find particularly good, and refer to recommended books.

Attracting animals
Again there are many good books which contain suggestions for attracting birds, mammals and invertebrates. But it is worth referring to the BUGS project web site because their research has found that some of the suggestions are a waste of time. One thing which they found successful was the provision of simple nesting sites for solitary bees, made from bundles of drinking straws. This is well worth doing, as these bees are important pollinators.

Hazards of which to beware when encouraging biodiversity

There is an adage 'Don't believe everything you read in the newspapers'. I'm going to suggest that the same applies to books (and now I am putting myself on the line!). *Be critical* and get into the habit of making your own observations.

I can give you some examples of occasions when I have followed books and found myself in trouble.

Pond weed

In lists of suitable floating-leaved plants for your pond you may see the broad-leaved pond weed (*Potamogeton natans*). I introduced it into my wildlife pond and bitterly regretted it. It has rhizomes which spread rapidly and are very tough; they soon form a thick mat on the bottom. When I came to try and remove some, it was incredibly difficult. Stick to the plants I've listed on page 124, which are easy to control.

Hawthorn, blackthorn and bramble

Hawthorn and blackthorn (sloe) appear in lists of shrubs suitable for hedges. But beware, they are *very* invasive and may crowd out more retiring species such as dogwood and guelder rose. Bramble is, in theory, valuable for its flowers which provide nectar for moths, and its berries as autumn food for birds, but its habit of spreading rapidly by suckers and layering can be very troublesome. It's not a desirable one to encourage.

Wildflowers in borders

I once thought it would be a splendid idea to introduce some ox-eye daisies (*Leucanthemum vulgare*) into my herbaceous border. I grew some from seed and planted them out. It was a disaster. What I hadn't realized was that, growing in a meadow, though flowering prettily, the straggly nature of their stems isn't visible. Exposed for all to see they looked embarrassingly unsuitable for a formal bed. This emphasized for me how horticultural cultivation over the centuries has bred out such features as untidy habit, short flowering period, small flower size, and has concentrated on the features that give the most handsome display.

But most of the advice in the literature is excellent, and very worthwhile following.

7. Genes, GM and the brave new world of designer plants

Genes pop up everywhere these days; hardly a week goes by without a report of scientists having discovered a new one, responsible for this or that characteristic. So it's worth having a look at what genes actually are, and what they do inside plants.

Introduction to DNA and genes

I'm going to return briefly to chemistry, because at the control centre of what goes on in plants and other living organisms is an amazing group of chemical compounds – the nucleic acids. The one most heard about is **deoxyribonucleic acid** – **DNA** – the stuff of which genes are made. The miracle is that the DNA inside the tiny nucleus (approximately one hundredth of a millimetre in diameter) of a single fertilized ovule holds enough information to direct the growth of a giant redwood tree. It does this (not as present-day computers do, using a binary system of on/off switches) in code form using four 'letters' arranged in groups of three (usually called **'triplets'**). The code directs the construction of proteins, many of which are the enzymes (see Chapters 1 and 3) which help along all the complex chemical reactions going on inside cells. So in the fertilized ovule of a baby plant, for example, certain genes will be sending out instructions to make the enzymes necessary for initiating cell division and the formation of a seed.

DNA structure and how it stores information

DNA structure is basically quite simple. A DNA molecule is made up of a very long double chain. Each chain is composed of thousands of units linked together like a string of beads. Each bead is composed of three components: ribose (a simple sugar not unlike glucose); a phosphate group (a few hydrogen and oxygen atoms combined with phosphorus – hence the importance of phosphorus as a nutrient); and a so-called 'base'. The only thing that varies along the chain is the nature of the bases. There are four distinct kinds, the foundation of the four-letter code:

- adenine (A)
- thymine (T)
- guanine (G)
- cytosine (C)

The convention is to use their initial letters. You may have seen on television, or in print, enormous charts with lists of these letters, repeated in various triplet combinations. Those are representations of DNA's secret code.

The other important feature of DNA is that the two single 'strings of beads' are linked together by the bases which have an attraction for each other – adenine latching on to thymine and guanine binding with cytosine. The structure is represented in very simplified form in Figure 7-1.

I haven't yet mentioned the term **'double helix'**. The double chain is actually coiled into a helix shape; imagine a flexible ladder twisted into a spiral. This feature was discovered by two Cambridge (England) scientists, James D. Watson and Francis H.C. Crick, who wrote an account of their work and the personal and social background. *The Double Helix* by James D. Watson is invaluable for its insights into how science operates; not only how the structure of complex molecules is worked out, but how human relationships influence the progress of an investigation.

The really fascinating thing about this amazing molecule is that the two strands of the helix can unwind and separate to allow new molecules of another similar type of substance, **ribonucleic acid**, **RNA**, to form on the surface of one of the exposed strands, the **DNA** acting as a template. Alternatively, the original DNA may replicate itself, as is necessary in cell division.

Another distinctive feature of DNA is its great stability, combined with the propensity for bits of the chain to break off and reinsert themselves elsewhere without upsetting the functioning of the molecule as a whole. Likewise, under laboratory conditions, bits of DNA can be inserted into a nucleus, and will find somewhere to slot themselves into the chain. This feature makes it possible to manipulate the genetic material – in other words to 'modify' it or 'engineer' it; hence the terms **genetic modification (GM)** and **genetic engineering**.

It now appears that Watson and Crick's original model, though having stood the test of time for over 50 years and probably reliable for most of the DNA in the nuclei of living cells, doesn't hold true in *all* cases. New discoveries are revealing that the double chain can perform some weird and wonderful tricks to form strange configurations – 'hairpin' shapes, folded zigzags, and even a triple helix. It is thought that some of these may be implicated in genetic diseases.

The code, a gene, and how proteins are made

Like DNA, proteins are built up of many smaller units, amino acids, linked together in linear fashion – beads in a string again. There are 20 important amino acids involved in the make-up of proteins, and each of the numerous proteins in plant

Figure 7-1. A short section of the DNA double chain. The helix is drawn unwound and shows, purely diagrammatically, how adenine always links with thymine and guanine with cytosine. The 'triplet' is a tiny part of a gene (see page 129).

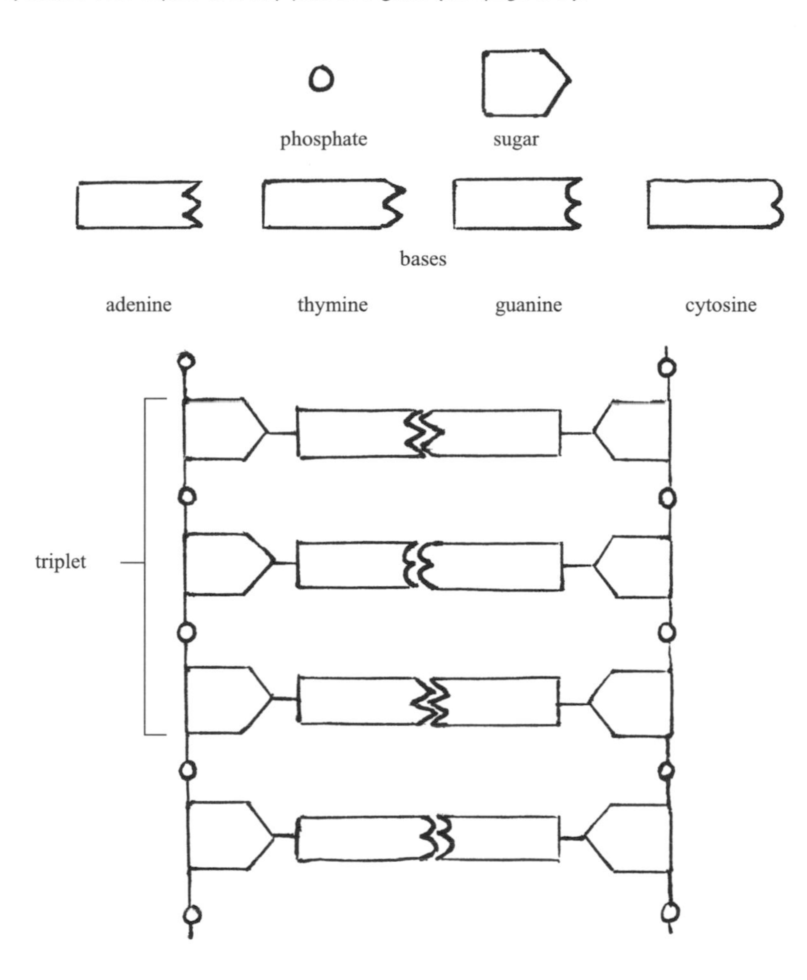

cells has a characteristic sequence. This sequence is determined by the **DNA code**.

The scientific work which has gone into unravelling the way in which the code operates is one of the great achievements of modern biology. It is currently understood to work as follows. Along the length of the DNA the bases A, T, G and C are arranged in triplets, for example, ACT or GTA and so on. Each of these triplets represents the information necessary for picking up one amino acid and fitting it into a protein chain. By a quite mind-bogglingly miraculous process, the strands of the double helix separate, and another molecule called **messenger RNA** forms itself on the template of one of the DNA strands, from smaller molecules floating around in the nucleus. This 'messenger', bearing the code, still in the form of triplets called **codons**, wanders off into the cell where it directs the formation of a protein. This piece of RNA represents **the length of DNA which carried the code for one protein**. And that stretch of DNA is a **gene.** Phew! – we've got there.

One of the chains carries the essential coded information and is called the **'sense strand'**. The double chain is drawn straightened out in Figure 7-1 but, in reality, it is coiled into a helix formation. One triplet, AGT, shown bracketed, carries the code for one amino acid. A gene is a long length of the chain bearing many triplets, carrying the code for the manufacture of a whole protein molecule.

How do genes determine plant characteristics and control processes?

By decades of experimenting in laboratories all over the world, biologists have established that DNA is responsible for directing the production of proteins. It's not so easy to work out the next stages, though. It's fairly straightforward understanding how one gene, responsible for the presence or absence of one protein, can determine colour, say, in the petals of a flower. If the protein is present it results in a purple pigment; if absent the petals are white. But that gene, if present, will be in *all* the cells of the plant; so why aren't the leaves and roots purple? No one understands the complete answers to such questions; but what biologists *do* know is that genes can be switched on and off – so genes for jolly colours will be switched *on* in flower petals but *off* in other parts of the plant. This is the simplest possible situation, but still leaves the question of how the switching on and off is done. Then there are far more complex processes to do with the growth and development of the different parts of a plant. One well-established fact is that there is a lot of interaction between genes themselves, and between proteins and genes. In other words, two genes acting together can produce an effect quite different from each one acting alone. Similarly the product of one gene can facilitate or inhibit the activity of another gene.

A great deal has been learnt about how genes control development by studying the development of flowers. A flower forms from a growing region (meristem, see Chapter 3) at the tip of a shoot. Figure 7-2 shows a section through a flower. Here three genes appear to be active which, in various combinations, determine the development of a ring of sepals outside, then petals inside the sepals, stamens next and, in the centre, the carpels of the ovary. How has this been worked out? The first step was to notice that in some double flowers, stamens or sepals are replaced by extra petals and, in certain species – for example the viridiflora rose – sepals replace all the floral parts. It was already known that sometimes genes undergo a spontaneous change called a **mutation**. A hypothesis could then be put

Figure 7-2. The ABC model of flower development.
(From Ingram, Vince-Prue and Gregory [eds.] *Science in the Garden* Royal Horticultural Society, 2002, by permission of Blackwell Publishing Ltd.)

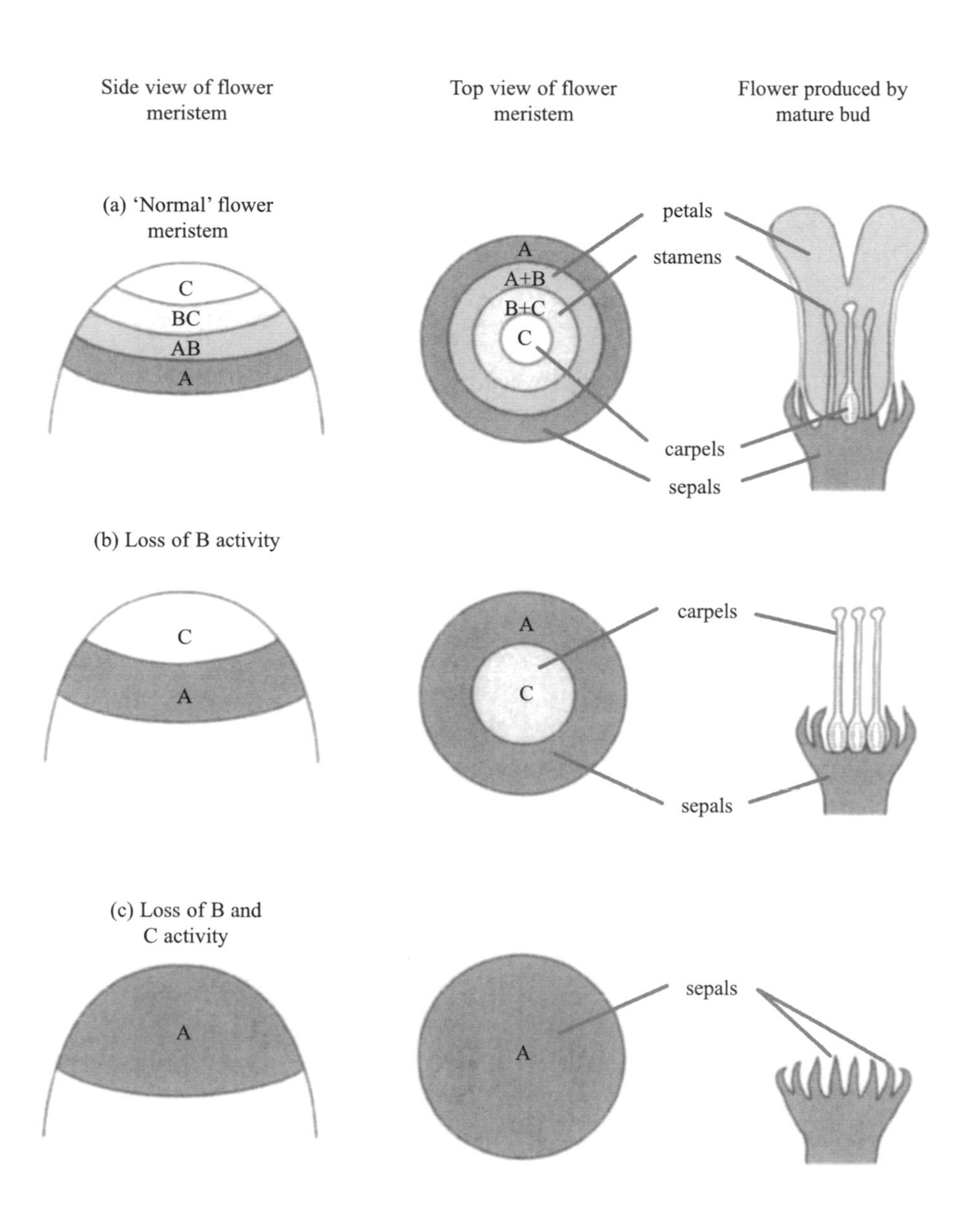

forward which suggested that there were genes in the cells of the shoot tip which, when functioning normally and switched on in various combinations, resulted in different flower parts developing but, when altered by mutation, produced the strange effects.

Testing this hypothesis with experiments on the mutant plants and comparing them with normal ones has resulted in a theory called the ABC model of flower development. There appear to be three genes which interact, labelled for convenience A, B and C. Gene A is most active around the edge of the meristem, while C is most active in the centre. B is active in a region between these zones which overlaps A and C. The protein product of A on its own produces sepals, C on its own produces ovary carpels, while B plus C results in stamens and B plus A gives rise to petals. (Figure 7-2 on page 133 shows that, in an experiment in which you could somehow knock out genes B and C, what would result would be all sepals.)

This arrangement of layers of cells, with the different layers controlled by particular genes, is turning out to be a general feature of plant development. The other kind of strange occurrence is the existence of variegated plants, or **chimeras**. The study of these has revealed that the pale parts are the result of a mutation occurring, during the early stages of cell division, in a gene that controls a layer of cells. This explains the striped effects, for example in *Hostas*, or the pattern in *Pelargoniums* where the centre of the leaves is pale and the periphery green.

All this leads on to the question of how our knowledge of the structure of DNA, and genes as discrete stretches of the DNA double helix, ties up with inheritance of characteristics; with Mendel and his pea experiments (see page 135) and plant breeding in general.

Chromosomes

In the nucleus of a cell the DNA doesn't just slosh about in its raw state; it is tightly coiled up and packaged with special proteins. Just before cell division or the formation of gametes these are visible under the microscope formed into sausage-shaped structures, the **chromosomes**. The total complement of chromosomes in a cell is called the **genome**. The other clever thing about the chromosomes is that they are arranged in **pairs**. The members of each pair carry the same set of genes. Those paired genes may, however, be exactly the same; *or* they may be very slightly different. In other words, a particular gene may exist in different forms each with slightly different effects. These forms of the same gene are called **alleles**. Let's call them **A** and **a**.

When pollen cells and ovules are being formed in the flower in preparation for reproduction, there has to be a *halving* of the chromosomes. If this didn't happen, there would be doubling up of chromosomes every time reproduction

Figure 7-3. All the possible combinations of gametes from the two parents.
If it so happens that **A** has a stronger effect than **a,** in a combination **Aa** only the effect of **A** will show; **A** is called the **dominant** allele of the gene while **a** is called **recessive**. The effect of **a** will only show up in the combination **aa**. In the above cross the effect of the recessive allele only shows up in one quarter of the offspring.

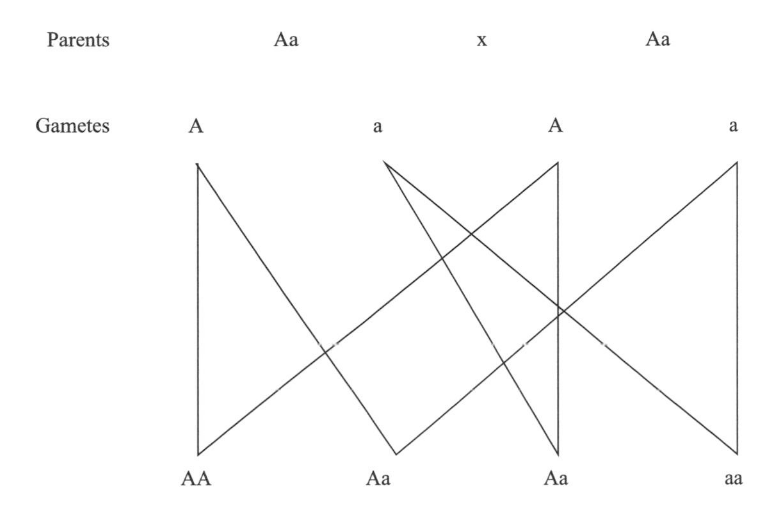

occurred (pollen grain nucleus fusing with ovule nucleus) which would, in most circumstances, be disastrous for the plant. So as the reproductive cells (**gametes**) are forming in a plant which possesses both **A** and **a** there is a separating out of alleles – one member of a pair of chromosomes with, say, allele **A**, is represented in some of the gametes and the other, allele **a,** in the remaining ones. When fertilization takes place there is then a random recombination of the alleles – **A** with **A**, **A** with **a**, or **a** with **a** to give the three possible arrangements – **AA, Aa** and **aa** (**aA** would be the same as **Aa**) (see Figure 7-3).

Simple plant genetics

Gregor Mendel was a monk, living and working during the mid 19th century in a monastery in what is now the Czech Republic. He was fascinated by the effects of crossing different varieties of plant and, in particular, noticed that certain cases of inheritance in peas followed a distinct pattern.

Mendelian crosses

He kept careful records of the results of his cross pollinating and concluded that there were discrete agents in the plants that carried characteristics independently from one generation to the next. A particular characteristic might disappear in the first generation of offspring, only to reappear in the next – a phenomenon familiar to us in human inheritance. (What we don't hear so much about is that there were cases where the simple, clear-cut effects I'm about to describe did *not* occur. Mendel chose to ignore those. To keep things simple, I shall also draw a veil over them!)

One of the pea varieties with which Mendel demonstrated the most marked effects were ones with wrinkled-skinned and smooth-skinned seeds. He developed pure-breeding lines of the two varieties – that is, however much crossing within each line the same offspring were always produced. He then crossed pure-breeding 'wrinkled' pea plants with a pure-breeding 'smooth' variety to give a first generation – called by geneticists the F1 generation. He found that all the offspring in the F1 had smooth peas (see Figure 7-4).

He then interbred F1 plants and the wrinkled ones reappeared in the proportions 3 smooth to 1 wrinkled. Figure 7-5 below shows the pattern. A similar pattern of inheritance was obtained with yellow and green peas, tall and dwarf plants, red and white flowers and one or two other features.

Let us suppose that **S** represents the dominant allele (version of the gene) giving **smooth** peas and **s** represents the recessive allele **wrinkled**. Remembering that chromosomes and the alleles of their genes occur in pairs, each individual will always have two alleles, which separate out into the gametes before fertilization takes place. A plant producing smooth peas could be either **SS** or **Ss.** One giving wrinkled peas could *only* be **ss**. (There are technical terms for the different combinations: where the alleles are the same, **SS** or **ss,** the plant is called **homozygous;** where they are different the term is **heterozygous**.)

What happens during reproduction can be shown as follows:

- In the F1 generation all the individuals have smooth seeds, but wrinkled ones reappear in the subsequent F2 generation. Geneticists call this **segregation** and **independent assortment**.
- If the plants with the wrinkled seeds are now crossed back with F1 plants (called a backcross) the offspring have smooth and wrinkled peas in equal proportions – Figure 7-6 (page 138).

The factors controlling the inheritance of the trait therefore appear to be independent units, which we now know are the alleles of one gene, a section of DNA, on a particular pair of chromosomes.

The fact that genes are arranged linearly on chromosomes was demonstrated in the early years of the 20th century by breeding experiments with fruit flies, tracing the fate of two pairs of characteristics rather than just one. The results showed that certain characteristics seemed to be linked closely together, some not

Figure 7-4. Mendel's F1 generation.

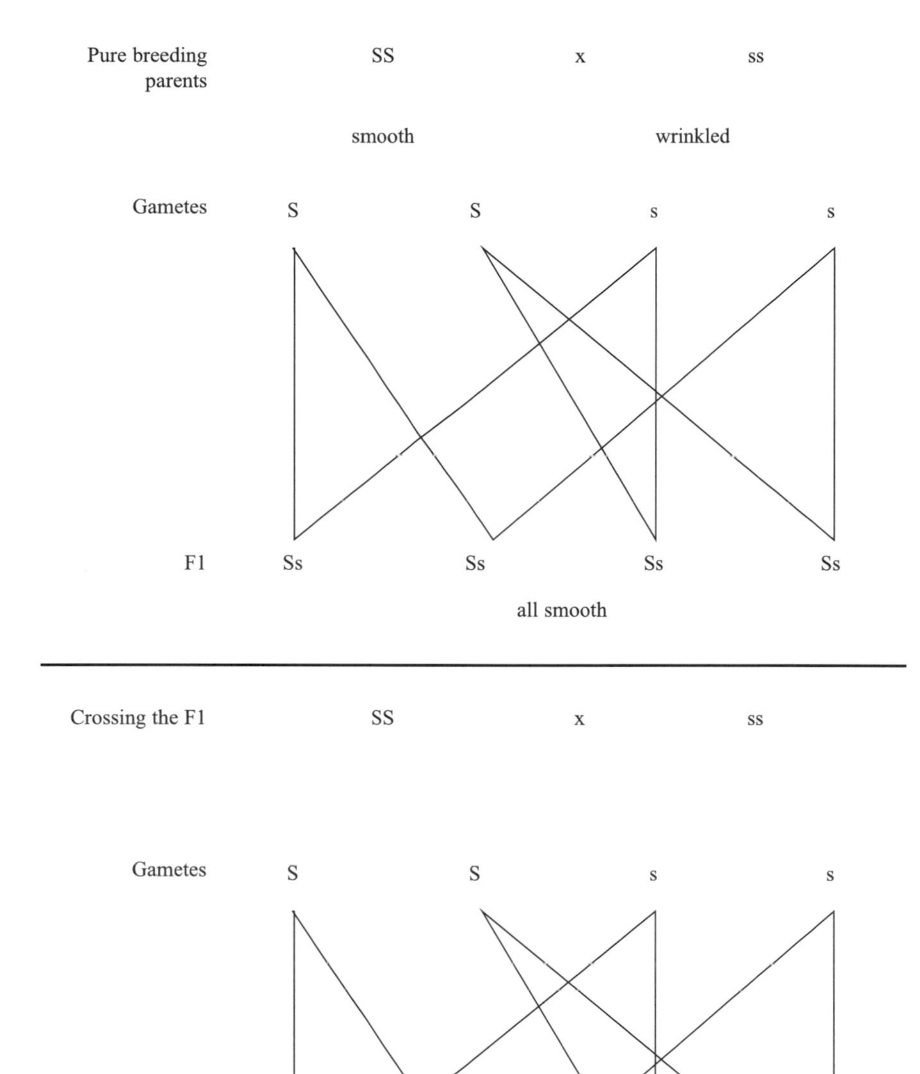

Figure 7-5. Mendel's F2 generation.

Figure 7-6. Smooth and wrinkled peas produced in equal proportions.

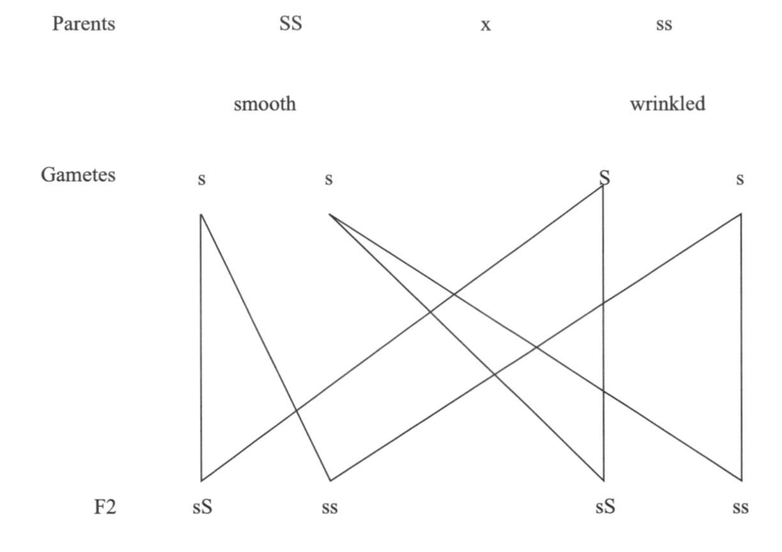

so closely and others not at all. A picture began to emerge of genes arranged linearly on rod-like structures, which could now be seen under the microscope in specially stained preparations of cell nuclei. These were the first indications of the structures we now know as chromosomes.

Cross-breeding between species

So far, everything described has applied to breeding *within* one species of plant, in other words, crossing varieties or cultivars. It is possible, however, to cross-breed species, though the resultant plants are usually sterile and subsequently have to be propagated vegetatively by grafting or other means (see Chapter 3).

Gardening implications

The F1 generation will not breed true

If you have seed which has come from an F1 plant, with flowers, say, which are all red, you will get a mixture of flower colours when you grow from that seed.

Hybrid vigour

For some reason not yet fully understood, the 'hybrids' of an F1 generation often have enhanced characteristics such as vigorous growth, sturdy habit, large flowers and so on.

They are therefore highly prized and, because they have resulted from careful breeding of pure lines and subsequent crossing, involving meticulous precautions against self-pollination, they tend to be more expensive than other varieties.

All is not so simple – origins of the enormous variation in plants

The simple and straightforward effects recorded by Mendel and other early geneticists are not very common. In reality, things are a good deal more complicated. This is due to factors, some already touched on, such as *interaction* between genes, the ability of bits of DNA to break off and *relocate* themselves (**chromosome mutation**) and the tendency for the code in a gene to change, however slightly (**gene mutation**). Another factor is that there is not always clear-cut dominance of one allele over another. This is termed **incomplete dominance** and its effect is that the individuals which have one dominant and one recessive allele (for example **Rr,** where **R** gives red flowers and **r** white), instead of being red are intermediate in colour – in this case pink.

It is because of all these complex factors that plants vary so much. In Chapter 1 I talked about the fact that when a packet of seed is sown it is most likely that there will be considerable differences between the resulting plants (see page 10). A lot of this variation is due to the genetic factors just described. Some of it will be due to factors in the environment but, even if the environment were to be made entirely uniform, for example, under laboratory hydroponic conditions, there would still be some variation due to genes.

Conventional plant breeding

This genetic variation is the stuff on which plant breeders work. In a batch of new plants there will be some with desirable traits and others with less desirable ones, be it flower shape or colour, attractive foliage or regular shaped fruit. The desirable ones are allowed to cross pollinate amongst themselves and the resultant seed sown. This is repeated for several generations, each time selecting the best plants. The final results of this kind of selection will be an open-pollinating cultivar.

Alternatively, the desirable plants may be inbred for several generations to give pure-breeding lines. These lines are then crossed to give F1 hybrid cultivars.

Doing your own breeding experiments is great fun, but ensuring the exact pollination you need is easier said than done. Transferring pollen from the stamens of one flower to the stigma of another is best done with a soft paint brush, but it's essential to make sure the brush is sterile and not contaminated by pollen from a previous experiment. Then, even in a greenhouse, there are insects flying around, so your experimental plants need to be isolated in some way.

Cloning and micro-propagation

Cloning simply means producing offspring, genetically identical to the parent organism, without sexual reproduction. It has been happening naturally in both plant and animal kingdoms since their very origins, but in plants and relatively primitive animals it happens more easily than in advanced vertebrate (backboned) animals. Because the more advanced animals don't have the ability to grow whole bodies from small bits it is extremely difficult to clone them. Very recently, however, there have been amazing advances in biotechnology which have made it possible to clone mammals and, controversially, human embryos. Plants, however, are much easier to manipulate, and plant cloning has been going on quietly in horticulture for centuries. Taking cuttings, and the many other forms of vegetative propagation (see Chapter 3) are, in essence, cloning. In nature some rhizomatous grasses have single clones covering very large areas.

More sophisticated methods are now used in specialized laboratories, using the capacity of plants to grow themselves from tiny fragments, under the right conditions and encouraged by the right hormones. The starting point is a small piece cut from a parent plant. The piece is often taken from a growing point where cells are being newly created and are usually free of viruses. To ensure sterility the cells may be taken from internal tissue, or sterilized before being place in a culture medium. The tiny fragments are first of all placed in a supporting gel which contains all the necessary nutrients for growth. These must include sugar, because the cells of the fragment may not initially contain chloroplasts and are unlikely to start photosynthesizing for some time. Hormones are also used (see Chapter 3). A high cytokinin/auxin ratio is first of all included to stimulate shoot growth, and later the cytokinin level is reduced to allow root formation to start. The baby plants are eventually transferred to soil.

Recombinant DNA technology – genetic modification

Artificial plant breeding has been going on for millennia. It can be argued that it began during the origins of agriculture, one centre of which is thought to have been in the Middle East about 12,000 years ago. Here the first farmers were domesticating wild animals and selecting hybrids between species of wild grasses which eventually came to be known as wheat.

Conventional breeding has been very successful in producing the vast variety of plants available today for our delight and nourishment. You might well ask, then, why it is necessary to bother with producing so-called 'GM' plants. In any case – what exactly is the difference between genetic modification and the juggling around of alleles of genes in conventional breeding programmes?

The answer to the first question is that genetic modification is potentially quicker and more reliable. The reason for this will be apparent, I hope, when I have answered the second question. What follows is an objective description of the techniques involved. The controversies over whether it is ethically and socially acceptable will be left to a separate section.

'Recombining' DNA – how are genes transferred from one plant to another?

The technology for engineering GM plants has been made possible by the discovery of three remarkable properties of the DNA molecule.

- DNA can fairly easily be extracted from the nuclei of cells and kept in laboratory conditions, where its code can be studied.
- It is extremely tolerant of bits being broken off its chain, and it has the capacity to repair the gap.
- If a short length of DNA, say a few dozen of the 'beads in the string' (see page 129), is introduced into a cell this bit can be made to 'home in' on a section of the DNA in a pair of chromosomes and slot itself into the chain.

These miraculous features of this extraordinary chemical substance make it possible for a gene to be identified, to be snipped out of the DNA chain and to be reintroduced into DNA from another source.

In living cells, fixing broken-off bits of DNA into place is carried out by special enzymes. Many of these enzymes have been identified and are kept in laboratories so that they can be used in experiments. Another useful thing which has been discovered is that a gene can be inserted into the genetic material of a bacterium, which can then be introduced into plant tissue, where it will convey the gene into the plant's own DNA. A similar trick can be accomplished using viruses. DNA can even be incorporated into a cell's genome by putting a short length into the culture medium in which protoplasts (plant cells without their cellulose cell walls) are floating, and then shooting an electric shock through the mixture.

Another important breakthrough has been the ability to unravel the complete DNA code of the genome of a plant, which makes possible an eventual understanding of the function of each and every gene. A few years ago it was announced that the code had been deciphered for the genetic material of the thale cress *Arabidopsis thaliana*, a rather insignificant plant, regarded by most gardeners as a weed. One fascinating thing to emerge has been evidence that there are certain genes which are common to many plants; some are possibly common to *all* plants. An example is a gene which is responsible for sensitivity to the hormone gibberellic acid (see Chapter 3). There is a mutated form of this gene, discovered in the thale cress, which causes insensitivity to gibberellic acid and therefore prevents cells elongating in the stem. The effect of this is to produce dwarf plants. It has been possible to snip out this gene and insert it into cells of rice plants. This would be done in cultures of small plant fragments, as in micropropagation (see page 57) and, lo and behold, the plants grown from these modified pieces were dwarfed. The same effect has been demonstrated in chrysanthemums and it looks, therefore, as though this is a gene which controls a very basic process in plant development, probably much the same in all plants.

It should be emphasized that the shooting of bits of DNA into plant cells is a hit-or-miss affair. It doesn't happen successfully in all the cells targeted, and there is a lot of failure and wastage.

Relevance of GM to gardeners

At present the technology involved in genetic engineering, of the kind just described, is extremely expensive. It is therefore unlikely that it will be used yet for producing plants to sell to ordinary gardeners. Its applications in the near future will be mainly in agriculture – where, already, new strains of rice, maize and beet are being produced – and in commercial vegetable growing.

GM compared with natural processes

The shuffling around of genetic material has been going on naturally since time immemorial. It happens during the formation of gametes, and their fusion in fertilization. There is, though, a difference between that and the artificial manipulation of genes which is not often highlighted. Remember that any one gene is represented by two alleles (two versions) on each of a pair of chromosomes (see page 134). In genetic modification, a new gene, when introduced into a plant cell, is able to incorporate itself *straight away* into *both* chromosomes of the pair. This is rather different from the sorting out and recombining of individual alleles in natural reproduction. Using capital and lower case letters for dominant and recessive alleles as before, consider a gene which has a dominant allele **T**, say, for tallness and a recessive allele **t** which codes for dwarf growth. **TT** and **Tt** would result in tall plants and **tt** in dwarf. At any event of fertilization between gametes (which possess only one allele, either **T** or **t**), recombination occurs in any of the three possible permutations. Suppose that a sturdier smaller variety is desired that will breed true and it is decided to obtain such a plant by genetic modification. The recessive form of the gene will be isolated in the laboratory and introduced by one of the methods described above. What will happen is that the form **t** will 'home in' on both the chromosomes, and insert itself into the DNA chain, to give **tt** and dwarf plants as desired. So a *whole* gene has been introduced into the genome in one fell swoop, instead of alleles rearranging themselves.

Another kind of reshuffling takes place when two species hybridize in the wild. This is a rare event in the animal world, though it has been found that birds kept in captivity and brought up together will interbreed freely. (At the Wildfowl Trust in Gloucestershire, England, there are specimens called Rarity Pies!) Defining a species, incidentally, is difficult. It is often described as an interbreeding group of organisms which cannot breed with any other group to give fertile offspring (see Chapter 8, page 147). But more latterly there are those who regard the species concept as a 'moveable feast' which doesn't really have any clear-cut definition.

In the plant world this is certainly more common. Two kinds of plant, considered to be separate species, may cross pollinate and produce offspring in which the chromosome number has become doubled up, or even trebled, to give a condition known as **polyploidy**. These individuals are fertile and become another species in their own right. An example of this is our modern wheat. There is good evidence that it has arisen as a result of this kind of hybridization between other primitive species of grass-like plants during the period of early farming in the Middle East and Europe. So, again, transference of genetic material to produce something quite new occurs in nature.

Another phenomenon which is very *similar* in both natural situations and GM is the introduction of viruses into genomes. It is well known that viruses 'make their living' by gaining access to the nucleus of a cell and taking over the DNA's own mechanism for replication. The virus reproduces at the cell's expense leaving mayhem and destruction behind it: hence viral disease. There is evidence also, though, that bits of virus DNA can get incorporated into the cell's DNA, to stay there indefinitely. It has been suggested that some major changes during evolution have been due to this kind of event, though finding ways of providing good scientific support for this is difficult. Bacteria are notorious in their ability to swap genetic material. It has been found that bits of DNA can be transferred from one bacterial cell to another with ease. So what geneticists term 'horizontal gene transfer' is quite common in the world of micro-organisms.

Based on what we know about natural gene transfer one can therefore argue that genetic engineering to produce GM plants is not that different from natural processes.

Ethical and safety considerations

As with the development of any new technology, worries about human welfare arise. For example, will genes introduced into one species spread to other species by natural hybridization, where they will have undesirable effects?

Particular cases in the UK at the time of writing are the concerns over a gene conferring resistance to herbicide in oil seed rape becoming transferred to closely related wild plants which may then become weeds which can't be controlled. Another concern is the effect on wildlife in the countryside, particularly where a gene conferring insecticidal properties is introduced. But maize has been produced in America which has had such a gene inserted. The gene results in the production of a toxin, harmless to humans, which kills a common pest. The alarm is over the question of transference of toxin-containing pollen to places where other, more desirable, insects could eat it. These are practical questions of safety and conservation. A more ethical consideration is the protection of the right of an organic farmer to be confident that his certified organic crops are not being contaminated with pollen from a GM crop. Bees can fly a very long way! Again, those who believe in a deity, who guides the course of life on earth, will perhaps argue that it isn't right to 'play God' with the genetic material.

What might be the benefits? There are potentially many. Some opponents of GM have argued that there is a danger that toxins may be introduced into the food chain. But, equally, the reverse may be possible. I have a daughter-in-law who has a serious nut allergy. It may be possible by genetic modification to *remove* the harmful proteins in certain nuts. In Asia there is already a variety of GM rice enriched with vitamin A which has obvious potential benefits. Scientists are also working on engineering crops which are able to secrete citric acid into the soil. This has the effect of making phosphate more readily available for absorption. In many developing countries salination of the soil after intensive irrigation is a serious problem, and salt-tolerant crops are therefore potentially highly desirable. If biotechnology can, without attendant harm, provide such plants, is it not ethically right?

Science *by itself* can't answer ethical questions, but decisions should always be guided by whatever help science can offer. Such help may be in the form of specially designed trials or experiments to test questions that arise. Unfortunately, because living organisms are so complex (see Chapter 1), the results of such trials may not provide clear-cut answers. Proponents of different points of view may then give the results different interpretations. Genuine scientists do their best. Their credibility within their profession depends on objectivity and honesty. But they are human and, from time to time, may make mistakes, or be influenced by the need to please their paymasters. But this shouldn't make us distrust science in general. As noted in the introduction, it is better than guesswork, anecdotal evidence or mere opinion.

Despite the slight difference between GM methods and natural reproductive recombination of alleles, it seems to me that genetic engineering is no different *in principle* from artificial breeding. There are many similarities between artificial and natural gene transfer. There has been always been some gene transfer between specially bred crop varieties growing in fields close enough together for bees to travel from one to the other. This has sometimes been a problem for farmers wishing to keep seed from a pure breeding strain. Planned mating or pollination, and subsequent selecting of desirable offspring, has been going on for a very long time. A dachshund dog, or double nasturtiums are, for me, very *unnatural* phenomena. Producing such animals and plants is surely just as much 'playing God' as isolating genes and introducing them into new organisms.

I have no objection to genetic engineering in principle. But I think that every individual case should be carefully and independently considered. Every possible attempt should be made to check that there will be no harmful effects, just as is done in testing drugs before they can be released for public use. There should also be safeguards against abuse of patented new cultivars, for example monopolies held on new crop varieties by large companies, and pricing of seed out of the reach of poor farmers in developing countries.

Making sure that GM is used for public benefit, rather than causing human ill health or environmental damage, won't be easy; things will surely go wrong from time to time. But, overall, there is no reason why it can't help us in the long run. And eventually its benefits may reach the domestic garden!

8. Plant classification and naming

Classification

Our prehistoric ancestors, observing the living things with which they shared their environment, would have noticed that there were certain obvious ways in which organisms can be grouped together, or divided up – whichever way you like to look at it. There were ones that were green and didn't move about and ones which were, on the whole, not green, and moved around to seek and catch food. They would have given them names – in modern English the former, 'plants', the latter, 'animals'. This would have been the beginning of **taxonomy**, the ordering of living things and the placing into groups according to our current knowledge of them.

Children practise taxonomy as they learn about the world around them. A child's early experience and primitive naming of things is often carried into adult life without the benefit of a more scientific perspective, so that the word 'animal' is used to refer to the sort of cuddly, furry creature with which children are most familiar. In other words, 'animals' are often what a biologist would call 'mammals', other animals (birds, fish, insects, and so on) not being counted as such. How often do you read in the press about 'animals, birds and insects' – as if the latter two were not animals at all.

Taxonomy is the practical side of the operation. Another term, **systematics**, is usually understood to refer to the more theoretical side in which *reasons* are sought for the groupings. Nowadays these reasons are to do with the study of evolution and how organisms are related to each other by virtue of their ancestry. Until the development of DNA analysis techniques, the working out of family trees was attempted on the basis of comparisons of structure and the study of fossils. Nowadays comparison of DNA is a key tool. But in taxonomy nothing is easy, and there are different schools of thought about many aspects of classification of living organisms, particularly when it comes to the larger groups.

The basic, smallest, unit of taxonomy is the species. There is something very 'real' about the concept of a species. A child has no problem distinguishing an English robin from a magpie – all robins have red breasts and can breed successfully with one another but not with magpies; all magpies look black and white, breed with each other but not with robins. Buttercups are very different from daisies and don't cross. The species is thus a basic biological entity, and this unit is at the bottom of the pile of taxonomic groups – or, to look at it another way, the very tips of the twigs of an evolutionary tree.

Unfortunately, in practice, the designation of a species is not always as straightforward as the robin and magpie or buttercup and daisy examples. For a start a child will be confused when it comes to blackbirds, because male and female blackbirds don't look the same. There will be other variations in appearance, some of which may depend on regional or other geographical factors. We then have to rely on the breeding definition – the ability of members of the same species to interbreed successfully to produce fertile offspring. Within that definition a category of sub-species can be recognized and, in horticulture, the variety or cultivar. There is even a term 'grex' which is given to a group of plants of the same parentage. But, in the plant world particularly (and to a certain extent in the animal world – see Chapter 7, page 143), there can be exceptions to the interbreeding definition because there can be hybridization between what were thought to be true species to give fertile offspring. There have also been cases observed where present-day species are undergoing change through time. Then there is also the problem of fossils where there may be a gradation of one form into another. So the species concept is not an entirely hard and fast one – it is the best that can be done for taxonomic purposes.

Species are gathered together into **Genera** (singular **Genus**), and Genera into **Families**, sometimes with subdivisions such as sub-species, and so on. After that we enter a minefield because botanists tend to name the larger groups differently from zoologists. There have also been major controversies over the largest groups, the Kingdoms. As more information has become available taxonomists have changed their minds. At the time of writing a generally accepted scheme is to divide all living organisms into five **Kingdoms** as follows:

Table 8-1. The five Kingdoms in the classification of living organisms.

Monera	This Kingdom includes the Bacteria, the Archaebacteria and the Blue-green Algae. These are all prokaryotic, which means that they do not have a true nucleus bounded by a membrane but the DNA is found loose in the cytoplasm. The following four Kingdoms are all eukaryotic – they have cells in which the DNA is contained within a true nucleus.
Protista	These are single-celled organisms, including the single-celled Algae, the Protozoa (tiny single-celled animals) and the slime moulds (strange fungus-like organisms which can wander about).

Fungi These are many-celled organisms which have no chlorophyll, so they have to rely on ready-made organic matter for food but, unlike animals, cannot move about. It includes the moulds, toadstools/mushrooms and yeasts.

Plantae The green plants which make their own food by photosynthesis. They include the many-celled Algae (including red and brown sea weeds which have a pigment similar to chlorophyll), liverworts and mosses, ferns, conifers and flowering plants.

Animalia Organisms which feed on other organic matter and move around to find and catch it. It includes all invertebrates (animals without backbones) and vertebrates (with backbones) – fish, amphibia, reptiles, birds and mammals.

The **Animalia** are then sub-divided into **Phyla** (singular **Phylum**), **Classes, Orders, Families, Genera** and species, sometimes with subdivisions of these.

To classify ourselves:

- We are animals, because we don't make our own food by photosynthesis as do plants; nor do we feed on dead organic matter by secreting enzymes into our surroundings as do fungi; nor are we primitive, single-celled creatures. Our position in the kingdom **Animalia** is on account of our ability to move around in search of food which we usually acquire by killing and eating plants, fungi or other animals.
- We then find a place in the Phylum **Chordata** because, during development, we have something called a notochord, and in the Sub-Phylum **Vertebrata** because we have a backbone.
- Our Class is **Mammalia** on account of our hair and our production of milk to suckle our young.
- Our Order is **Primates** along with monkeys and apes.
- We belong to the Family **Hominidae**, the Genus *Homo*, and the species *sapiens* (Latin for 'wise' – though you may have reservations about this!) – hence *Homo sapiens*.

Plant classification has different terminology, which is most confusing. The term Phylum generally isn't used, botanists preferring to use the term Division. So we have the following **Divisions**:

- **Algae** (filamentous 'blanket weeds' and the sea weeds)
- **Bryophyta** (mosses and liverworts)

- **Pteridophyta** (ferns, clubmosses and horsetails)
- **Coniferopsid**a (formerly called Gymnosperms to include the familiar conifers – ginkgoes, cycads and yews have their own divisions)
- **Magnoliaceae** (formerly called the Angiosperms, to include all the flowering plants, within which are the **Monocotyledones** and the **Dicotyledones**).

To classify a meadow buttercup – it belongs to the Kingdom **Plantae** on account of its ability to photosynthesize, the Division **Magnoliophyta** because it has flowers with ovules protected inside an ovary, the Sub-Division **Dicotyledones** because it has branching-veined leaves, and two seed leaves. It is then placed in the Family **Ranunculaceae**, the Genus *Ranunculus* and the species *acris.*

Bar coding

Recently there has been a great deal of research into the bar coding of species as a way of identifying them. It is possible to do this in much the same way as items in a supermarket are identified. There is a call for the whole of the living world to eventually be treated in this way. It is hoped that this will encourage people to find out more about their environment and to care more for its biodiversity. At the time of writing this is being developed; keep an eye out for progress in this field.

The place of viruses and their differences from bacteria

So where do viruses fit in? The reason that they are not included in any of the five Kingdoms is that they cannot live on their own, independently of other living cells. In order to reproduce they are obliged to take over enzyme systems belonging to another living organism; in other words, they are *forced* to be parasites. They are therefore on the borderline of living and non-living.

Viruses are smaller than bacteria. They cannot be seen with a light microscope but can be studied with an electron microscope. This technique has shown that they consist of a 'package' of DNA or RNA surrounded by a coat of protein which often has a rather pronounced symmetrical shape. Unlike bacteria they don't contain cytoplasm or their own systems of enzymes – they are just tiny packages of genetic material which reproduces itself at the expense of other living things. During the process of reproducing they often cause fatal damage to their host cells, which is why they are often agents of disease.

Evolution and classification

Despite occasional spats between evolutionary biologists as to the way in which it has actually *happened*, it is generally accepted that living organisms have undergone change. Three and a half billion years ago it appears that there may only have been a 'soup' of complex organic molecules; then simple single-celled organisms appeared in this environment, followed by many-celled, but primitive, plants, fungi and animals to produce the multitudinous variety of living things we have today. If the evidence is examined dispassionately there can be no doubt about this, despite the strange refusal to accept the testimony, displayed by 'creationists', who claim that there has been no such evolution. The clearest evidence comes from the fossil record where an organism can be seen changing with time, a classic example being the horse which shows change from a dog-like creature to our present-day *Equus.* Such time-sequences are well-founded, since it is now possible to date the rocks in which fossils are found. There are gaps, and not all organisms show up as fossils, but that is not surprising considering that fossilization requires very special conditions which inhibit the natural process of decomposition.

Another line of evidence is observation of present-day change in a species. For example one of the Galapagos Island finches of Darwinian fame (the ground finch *Geospiza fortis)* has, since 1977, changed from being a small bird with a small beak to a much larger bird with a large beak. In 1977 a severe drought reduced the availability of seeds, the finches' food. The only birds to survive were ones with large beaks which could make use of larger items of food; so natural selection has resulted in a change in the species. Other evidence now comes from molecular genetics – DNA from apparently related species can be compared, rates of mutation established, and thus the time in the past at which the species might have diverged from a common ancestor.

Following all these lines of evidence our grouping of living organisms into categories can be tied up with evolutionary relationships. Species, genera, families and so on can thus be organized into genealogical trees.

Naming individual species

I used to explain to my students that Latin names were essential to avoid the confusion sometimes thrown up by the use of different common names in different regions of the country. This principle applies equally to different parts of the world, and is perhaps even more important today in our globalized telecommunicating society. An example of this confusion crops up in the most erudite of circles. The wild plant *Cardamine pratensis* has two or three common names, the most frequently used being lady's smock or cuckoo flower. There is also a plant, the wild arum, commonly called lords and ladies or cuckoo pint, the Latin name of which is *Arum maculatum.* In a recent delightful and learned book about the garden of a naturalist, there is a mix-up because the pretty cuckoo flower, food of the orange tip butterfly, is named as *Arum maculatum.*

The irony is that the scientific organization which authorizes the naming of plants is for ever revising things, so that both plant species names, and classification into families and other groupings, change in almost as confusing a way as the variable use of common names. The poor bluebell has gone through as many as six changes of Latin name since Linnaeus's time (see below). Quite recently major changes have been made to plant families so that, for example, the group formerly known as the Compositae (the dandelion and daisy-like flowers) is now called the Asteraceae, and the Trilliaceae and Amaryllidaceae, once families in their own right, are now included in the Liliaceae.

Be that as it may, Latin names, accompanied by a plethora of additional descriptors for cultivated varieties, are the convention for garden plants. The system was originated by the Swedish biologist Linnaeus (1707–78), who saw the need for some way of avoiding confusion when one of the main preoccupations of 18th-century naturalists was discovering new species and naming them. The system is called the **Linnean binomial (two-name) system –** a Genus name beginning with a capital letter, followed by a species name all in lower case. This is usually written in italics, or underlined. The variety name is then tagged on, without underlining or italics but with single quotation marks. So, roses belong to the Genus *Rosa*, within which there are about 250 recognized species and many varieties within those. An example might be *Rosa gigantea* 'Gloire de Dijon'. Sometimes a hybrid between two species is given a particular species name of its own, in which case it is preceded by x. So, a rose example might be *Rosa* x *odorata*, the cross between *Rosa gigantea* and *Rosa chinensis*.

To the best of my knowledge this is the current practice – but don't be surprised if things change! Classification and naming evolves, just as do living organisms.

Appendices

Appendix 1

List of small trees for woodland edge habitat

Large trees such as ash and beech have not been listed, but if you have a very large garden by all means include them. Blackthorn, hawthorn and bramble are good for berries and flowers but are very invasive – only feasible if you are very energetic or have a good gardener.

Alder	*Alnus glutinosa*	damp – acid (coppice)
Alder buckthorn	*Frangula alnus*	damp – acid
Birch	*Betula pendula*	dry – acid if possible
Bird cherry	*Prunus padus*	fertile – alkaline if possible
Crab apple	*Malus sylvestris*	
Dog rose	*Rosa canina*	
Elderberry	*Sambucus nigra*	
Field maple	*Acer campestre*	clay if possible
Gean	*Prunus avium*	fertile – alkaline if possible
Guelder rose	*Viburnum opulus*	
Goat willow	*Salix caprea*	damp if possible
Hazel	*Corylus avellana*	
Holly	*Ilex aquifolium*	
Oak (pedunculate)	*Quercus robur*	heavy if possible (coppice)
Oak (sessile)	*Quercus petraea*	light if possible (coppice)
Rowan	*Sorbus aucuparia*	light
Sallow	*Salix cinerea*	damp if possible
Spindle	*Euonymus europaeus*	alkaline
Whitebeam	*Sorbus* aria	
Wild privet	*Ligustrum vulgare*	

A few suggestions for other shrubs and conifers which may provide flowers for insects, berries for bird food, or cover in winter.

Buddleja	*Buddleja* spp.
Cotoneaster	*Cotoneaster* spp.
Cupressus	*Cupressus* spp.
Flowering currant	*Ribes sanguineum*
Ivy	*Hedera helix*
Juniper	*Juniper* spp.
Mahonia	*Mahonia aquifolium*
Pyracantha	*Pyracantha* spp.
Quince	*Cydonia oblongata*
Skimmia	*Skimmia* spp.

Appendix 2

Suggestions for plants in ground layer of woodland edge

Spring	
Common dog violet	*Viola riviniana*
Daffodil	*Narcissus pseudonarcissus*
Ransoms	*Allium ursinum*
Snowdrop	*Galanthus nivalis*
Sweet violet	*Viola odorata*
Wood anemone	*Anemone nemorosa*
Wood sorrel	*Oxalis acetellosa*

Summer	
Bluebell	*Hyacinthoides non-scripta*
Foxglove	*Digitalis purpurea*
Greater stitchwort	*Stellaria holostea*
Herb Robert	*Geranium robertianum*
Red campion	*Silene dioica*
Yellow archangel	*Lamiastrum galeobdolon*

Appendix 3

Suggestions for plants in early and late meadows

Lists vary enormously; refer to other books as well. Check growing requirements if necessary, particularly if you have dry chalky soil as other species may be more suitable.

Early summer

Birds foot trefoil	*Lotus corniculatus*
Black medick	*Medicago lupulina*
Bulbous buttercup	*Ranunculus bulbosus*
Cats ear	*Hypochaeris radicata*
Common vetch	*Vicia sativa*
Cowslip	*Primula veris*
Daisy	*Bellis perennis*
Dandelion	*Taraxacum officinale*
Lady's smock	*Cardamine pratensis*
Meadow vetchling	*Lathyrus pratensis*
Mouse-eared hawkweed	*Hieracium pilosella*
Ox-eye daisy	*Leucanthemum vulgare*
Pignut	*Conopodium majus*
Red clover	*Trifolium pratense*
Salad burnet	*Sanguisorba minor*

Late summer

Field scabious	*Knautia arvensis*
Hardheads	*Centaurea nigra*
Meadow buttercup	*Ranunculus acris*
Perforate St John's wort	*Hypericum perforatum*
Rough hawkbit	*Leontodon hispidus*
White clover	*Trifolium repens*
Yarrow	*Achillea millefolium*
Yellow rattle	*Rhinanthus minor*

Appendix 4

Suggestions for areas of permanent grassland to represent a lightly grazed area, for example, in your lawn

Black medick	*Medicago lupulina*
Daisy	*Bellis perennis*
Mouse-eared hawkweed	*Pilosella officinarum*
Ribwort plantain	*Plantago lanceolata*
Self heal	*Prunella vulgaris*
Slender speedwell	*Veronica filiformis*
Tormentil	*Potentilla erecta*
White clover	*Trifolium repens*
Yarrow	*Achillea millefolium*

Appendix 5

Suggestions for a patch of colourful cornfield weeds

Chamomile	*Chamaemelum nobile*
Corncockle	*Agrostemma githago*
Cornflower	*Centaurea cyanus*
Corn marigold	*Chrysanthemum segetum*
Poppy	*Papaver rhoeas*
Scarlet pimpernel	*Anagallis arvensis*
Scented mayweed	*Matricaria recutita*
Scentless mayweed	*Matricaria perforata*

Appendix 6

Plants for the pond

Submerged plants for habitat and for oxygenation during the day

Hornwort	*Ceratophyllum demersum*
Water milfoil	*Myriophyllum spicatum*
Water starwort	*Callitriche* spp.

Floating-leaved plants

Amphibious bistort	*Polygonum amphibium*
Fringed water lily	*Nymphaea peltata*
Water crowfoot	*Ranunculus aquatilis*
White water lily	*Nymphaea alba*
Yellow water lily	*Nuphar lutea*

Free-floating plants

Frogbit	*Hydrocharis morsus-ranae*
Water soldier	*Stratioites aloides*
Water violet	*Hottonia palustris*

Emergent plants in deepish water

Bog bean	*Menyanthes trifoliata*
Bur reed	*Sparganium erectum*
Flowering rush	*Butomus umbellatus*

Emergent plants in shallow water or marshy ground

Brooklime	*Veronica beccabunga*
Great hairy willowherb	*Epilobium hirsutum*
Lady's smock	*Cardamine pratensis*
Lesser spearwort	*Rananunculus flammula*
Marsh marigold	*Caltha palustris*
Meadowsweet	*Filipendula ulmaria*
Purple loosestrife	*Lythrum salicaria*
Ragged robin	*Lychnis flos-cuculi*
Water mint	*Mentha aquatica*
Water plantain	*Alisma plantago-aquatica*
Yellow iris	*Iris pseudacorus*
Yellow loosestrife	*Lysimachia vulgaris*

Appendix 7

Some suggestions for formal beds

Lists vary tremendously, but these are some ideas for attracting adult insects (mainly bees, butterflies and moths) to feed on nectar or pollen. You could leave stalks for insects to hibernate in, for example ladybirds and lacewings (though they are carnivorous). Some birds, for example greenfinch, goldfinch or sparrow will feed on seed heads you leave; suggestions in other books are always useful. Where it says spp. be sure to get cultivars which have nectar and pollen.

Spring

Arabis	*Arabis* spp.
Aubretia	*Aubretia* spp.
Forget-me-not	*Myosotis scorpioides*
Honesty	*Lunaria annua*
Poached egg plant	*Limnanthes douglasii*
Polyanthus	*Primula* spp.
Sweet violet	*Viola odorata*
Wallflower	*Cheirantus* spp.
Yellow alyssum	*Alyssum* spp.

Early summer

Catmint	*Nepeta cataria*
Evening primrose	*Oenothera odorata*
Lavender	*Lavendula* spp.
Perennial cornflower	*Centaurea montana*
Phlox	*Phlox paniculata*
Pinks	*Dianthus* spp.
Rosemary	*Rosmarinus officinalis*
Sweet bergamot	*Monarda didyma*
Sweet William	*Dianthus barbatus*
Thrift	*Armeria maritime*
Thyme	*Thymus vulgaris*
Tobacco plant	*Nicotiana* spp.
Verbena	*Verbena* spp.

Late summer

Golden rod	*Solidago* spp.
Hyssop	*Hyssopus officinalis*
Ice plant	*Sedum spectabile*
Marjoram	*Origanum vulgare*
Michaelmas daisy	*Aster* spp.

Glossary

allele One of the alternative forms of a gene.
amino acid An organic compound from which proteins are constructed, a protein being made up of a chain of amino acids joined together.
angiosperm A plant belonging to the group of flowering plants now classified with the Division Magnoliaceae.
animal A living organism which has the ability to move to find food, lacks chlorophyll, and does not possess cell walls outside the cell membrane. The animal kingdom includes – amongst others – birds, amphibians, reptiles, fish, insects and all other invertebrates.
annual A plant which completes its life cycle within one year – growing to maturity, flowering, setting seed and then dying.
anther The part of a stamen in which the pollen is produced.
atom The smallest possible particle of an element that can be obtained by ordinary physical or chemical means.
auxin A plant hormone which acts to increase the size of cells and therefore promotes growth.
bacteria Micro-organisms which consist usually of single cells without true nuclei – the DNA being loose in the cytoplasm in a loop formation.
bare-rooted This usually refers to a young shrub or tree which has not been grown in a container and therefore has an exposed root system.
base A chemical substance which, in genetics, is the term given to one of the sub-units of the DNA molecule.
biennial A plant which takes two years to complete its life cycle – to grow to maturity and set seed.
biodiversity The variety of life – to include the variety of species and the variation within species.
biosphere That part of our planet made up of living organisms.
bog A type of vegetation consisting of plants adapted to living in acid peat.
bulb A plant storage organ constructed of the swollen bases of leaves.
calcicole A plant which is adapted to living in alkaline conditions.
calcifuge A plant which cannot tolerate alkaline conditions.
canopy The uppermost branches and leaves of forest trees.
carbohydrate A compound made up of carbon, hydrogen and oxygen in the proportions 1:2:1, found in the cells of living organisms and initially manufactured by photosynthesis.
carnivore An animal which feeds by catching and killing other living animals.

carpel A compartment of an ovary. Some ovaries consist of a single carpel containing either one ovule or several, while others have several carpels with many ovules in each.
carr Woodland growing in waterlogged conditions, usually composed of alder trees.
cell The basic structural unit of which living organisms are built up.
cellulose A structural carbohydrate which forms the cell walls of plants.
chlorophyll The green pigment in plant cells which facilitates photosynthesis.
chloroplast Flat oval structures in plant cells containing chlorophyll.
chromosome A thread-like structure in the nucleus of a cell in which the DNA is coiled and bound up with protein, visible as rod-like structures during cell division.
climax The final stage of ecological succession.
clone Offspring of an organism which is genetically identical with the parent.
codon A set of three bases in the DNA or messenger RNA molecule which carries the code for a single amino acid.
compensation point The point at which the rate of photosynthesis exactly equals the rate of respiration and there is no gain or loss of carbohydrate.
compost Decomposed organic material. The term may refer to material used as a fertilizer (soil improver) or as a growing medium, in which case it is often mixed with mineral material.
compound A chemical substance made up of two or more elements joined together.
coppice Used as a verb this means to cut the trunk of a tree near its base. New stems sprout from the base, or stool, which can be used as poles or other small pieces of wood. The term may be used as a noun to mean a small area of woodland or scrub.
corm A plant storage organ formed from the swollen base of a stem.
cultivar A horticultural variety which has been bred from the original wild species and maintained by cultivation.
cytoplasm The internal material of cells, contained within a cell membrane.
detrivore An animal which feeds on small particles of dead organic matter.
dicotyledon A plant belonging to the large group, Dicotyledones, in which all members have two cotyledons (seed leaves) in the seed. The mature plants have leaves with branching veins and a great variety of shapes.
differentially permeable membrane A membrane surrounding the cytoplasm of a cell which is permeable to certain small molecules but not to larger ones.
DNA Deoxyribonucleic acid – the chemical substance which carries the genetic information in code form.
dominant In genetics this refers to the allele of a gene which is always expressed and which masks the expression of the recessive allele. In ecology it refers to the species in a community which is the most abundant and is dominant over all the others.
ecological niche The way of life of a species – how it feeds, breeds and protects itself from predators.

ecosystem A network of living things interacting with the non-living components of their environment.
element A chemical substance that cannot be converted by chemical means into any simpler substance.
endosperm The food store in a seed.
enzyme A specialized protein found in living cells which acts as a catalyst for (facilitates) biochemical reactions.
etiolation Abnormally rapid growth of stems in the absence of light, stems and leaves lacking chlorophyll and appearing yellowish.
F1 The first filial generation of a cross between parent gametes.
fat An organic compound made up of carbon, hydrogen and oxygen, occurring naturally in plants and animals, solid at room temperature and insoluble in water. Related to oils which are liquid at room temperature. Collectively they are termed lipids.
fen Marshy area in which the plants are growing in neutral to alkaline peat.
fertilization The union of male and female gametes to form a zygote which subsequently may develop into an embryo. Also refers to adding extra nutrients to soil in order to improve productivity.
fertilizer A substance which can be added to soil to increase nutrients and improve productivity.
fungi Living organisms which lack chlorophyll but resemble plants in that they are unable to move. They feed by exuding enzymes which digest surrounding dead organic material into a form which they can absorb, or by parasitic means, feeding on the fluids in a living host.
gamete A reproductive cell or nucleus – in animals a male sperm cell or female ovum (egg cell); in plants the nucleus in a pollen grain which fuses with the ovum in the ovule during fertilization.
gene A stretch of the DNA molecule which carries the code for a single protein.
genome The total genetic material of an organism – the complete set of chromosomes plus any other DNA.
greenhouse effect The warming of the planet as a result of the trapping of heat by a layer of gases in the atmosphere. Solar radiation is converted to heat when it hits the ground but the heat is unable to return to space because it is absorbed by the 'greenhouse gases'.
Gymnosperm A plant belonging to the non-flowering group whose members have ovules borne unprotected, usually at the base of a scale of a cone.
habitat Strictly the place where a species lives, but often used loosely to mean a particular kind of plant community.
herbivore An animal that feeds on plants.
heterozygous An organism bearing two different alleles of a gene.

homozygous An organism in which a pair of alleles are identical.
hydroponics A system of cultivation in which plants are grown in a solution of nutrients.
hydrosere The effect of ecological succession in water bodies.
hyphae The thread-like structures which form the main part of a fungus.
hypothesis A statement, based on initial observations, ready to be tested for its validity by further observation or experiment.
invertebrate An animal without a backbone.
ion An electrically charged particle of an element or part of a compound. Cations are positively charged, anions are negative.
IPM Integrated Pest Management – a system of pest control which uses both artificial chemical and biological methods.
lignin A structural carbohydrate which is used to strengthen the walls of xylem vessels in the stems of plants.
lipid A general term for a fat or oil.
mainstorey The part of a woodland which lies between the topmost canopy and the shrubs of the understorey.
manure The faeces of farm animals mixed with straw and used as a fertilizer.
marsh A general term for an area of land which is frequently flooded and has soil composed mainly of mineral matter rather than peat.
meristem An area of a plant which is actively growing by the division of cells – found at the tips of stems and roots and in the cylinder of cambium inside the stems of woody plants.
micro-propagation Propagation of plants by taking a tiny portion of tissue and culturing it with nutrients and hormones so that it grows into a young plant.
mire A plant community growing in waterlogged peat.
mitochondria Tiny structures inside cells where the chemical reactions of respiration take place.
molecule The smallest unit of a chemical element or compound which can exist independently and take part in a chemical reaction.
monocotyledon A plant belonging to the group Monocotyledones in which the members have a single cotyledon in the seed. The leaf veins are parallel.
muck A popular term describing manure or compost.
mulch A covering placed over the soil to suppress weeds and retain moisture.
mutation A change in the genetic material which may occur spontaneously or may be induced artificially by radiation or chemical treatment.
mycelium The network or mat of fungal threads (hyphae) which forms the main part of a fungus.
mycorrhiza An association between a fungus and the roots of a plant which usually benefits both thc fungus and thc plant.

nucleus The structure inside a cell which contains the genetic material and acts as the control centre for the functioning of the cell.
organelle A tiny object inside a cell which has some kind of function. Examples are the mitochondria where respiration takes place and chloroplasts, responsible for photosynthesis.
organic Relating to living organisms or, if applied to chemical substances, those which are based on carbon atoms. Used in relation to gardening and farming it refers to cultivation without the use of man-made artificial chemicals.
osmosis The movement of water through the differentially permeable membranes of cells from a region where there are a lot of water molecules (a dilute solution) to a region where there are fewer water molecules (a more concentrated solution).
ovary A structure which produces and protects female reproductive cells. In plants it is only the Angiosperms (flowering plants) which have ovaries containing ovules.
ovule The structure which contains the female reproductive cell (gamete) and develops into a seed after fertilization.
ovum The actual female reproductive cell, the nucleus of which fuses with the male gamete during fertilization.
parasite An organism which lives on or inside another living organism, the host, and feeds from its host's tissues.
perennial A plant which continues to live for several years, either having woody stems, or dying back each winter but regrowing in the spring.
permaculture A system of horticulture in which plants are grown in a self-sustaining woodland kind of community with no major disturbance of ground.
pesticide Any chemical substance which kills living organisms likely to be pests in horticulture or agriculture.
pH A scale on which acidity or alkalinity is measured.
pheromone A chemical substance produced by animals as a signalling device – very often to attract a mate.
phloem A system of vessels in the stems of plants which are specialized for carrying organic compounds – mainly sugars from the leaves down to other parts of the plant.
photosynthesis The process by which plants manufacture sugar from carbon dioxide and water in the presence of chlorophyll, or a related pigment, and sunlight.
phytochrome A light-sensitive substance found in plants which helps the plant to respond to changing light intensity and day length.
pollination The process of transference of pollen from the stamens to the stigma of a plant or, in the case of gymnosperms, to the naked ovule.
polyploidy A condition in a plant in which the chromosome number has increased to three, four or more sets beyond the normal two.
predator An animal which feeds on other animals by pursuing and catching them.

primary woodland Woodland on a site which has been continuously wooded throughout historic times and therefore assumed to be continuous with the original natural ancient woodland.
protein An organic compound containing carbon, hydrogen, oxygen, nitrogen, and sometimes other elements. A protein molecule is very large and is made up of amino acids linked together into a chain, which may then be coiled or folded in complex ways.
pure breeding An organism which, when bred with another similar one, produces offspring identical with the parents. Members of this first generation when bred together also continue to produce identical offspring and this continues into subsequent generations.
recessive Refers to the allele (version) of a gene whose effect is masked by the other member of the pair, the dominant allele, when present together in an organism. It can only show its effect when two recessive alleles are present together on the pair of chromosomes.
respiration The process in which energy is released from glucose in living cells. The process in plants requires oxygen. In animals and some bacteria it can take place without oxygen, in which case it is termed anaerobic respiration.
rhizome A plant storage organ formed from an elongated swollen underground stem, usually growing horizontally near the surface.
RNA Ribonucleic acid – related to DNA and possessing a similar structure. Messenger RNA acts to transfer the genetic code from the nucleus to sites of protein synthesis in the cytoplasm of cells. Transfer RNA helps to carry out the construction of protein molecules. In some viruses RNA may be the main genetic material.
salt In chemistry this refers to the compound which results from the combination of an acid with a base. An example is sodium chloride (common salt) formed by the reaction of sodium hydroxide with hydrochloric acid. Others are potassium nitrate, ammonium sulphate, and so on.
secondary woodland Woodland which has been planted or colonized naturally on land which has previously been converted to another use, for example, pasture or arable.
species A difficult concept to define. The usual definition is a group of individuals which can interbreed to produce fertile offspring. But in the plant world there are exceptions – some groupings which have been thought of as species have been able to hybridize and their offspring have been able to breed, for example modern wheat is thought to have originated like this. Even in the animal world species change and evolve so that the exact delineation of a new species is sometimes uncertain.
stamen The male part of a flower which bears the anthers in which pollen is produced.
stigma The part of the female structure in the flower which receives pollen. This is usually at the tip of the stigma which is a stalk-like projection from the ovary.
stomata Pores in the outer layer of a leaf which allow the exchange of gases and the loss of water vapour.

style The stalk-like projection from the ovary of a flower.
succession The process whereby one plant community is replaced by another in a series of steps from bare rock through to the climax, which is usually woodland.
swamp A general term for a waterlogged area of land.
symbiosis An association between two species in which there is mutual benefit.
transpiration The physical process of loss of water by evaporation from the exposed surfaces of a plant, mainly from the leaves.
understorey The layer in a woodland which is composed of small trees not more than 10 m high.
vector An organism which carries a disease-causing virus or bacterium from one plant to another, for example, aphids can carry virus diseases – sucking them up from one plant as they feed and injecting them into another.
vertebrate An animal with a backbone.
virus An organism consisting only of genetic material (DNA or RNA) and a protein coat. They can only reproduce by taking over the enzymes of a living cell so cannot said to be truly living themselves but are on the borderline of living and non-living.
whip A young shrub or tree – usually not more than one year old.
xerophyte A plant which is adapted to living in dry conditions by possessing features which reduce water loss.
xylem The system of vessels in stems which conduct water up from the roots to the leaves.

Further reading

The following titles relate specifically to the chapters under which they are listed.

Chapter 1

Jensen, William A. and Kavaljian, Leroy G. (series eds.) *Plant Physiology* ch. 10, 11 and 12. Belmont, California: Wadsworth Publishing Company Inc, 1969.

Soper (ed.) *Biological Science* Vol 1 ch. 5, 6 and 7. Cambridge University Press, 1984.

Chapter 2

Jensen, William A. and Kavaljian, Leroy G. (series eds.) *Plant Physiology*. Belmont, California: Wadsworth Publishing Company Inc, 1969.

Russell, Sharman Apt. *Anatomy of a Rose: Exploring the Secret Life of Flowers.* Perseus, 2001.

Chapter 3

Ingram, Vince-Prue and Gregory (eds.) *Science and the Garden* ch. 6. Royal Horticultural Society, 2002.

Jensen, William A. and Kavaljian, Leroy G. (series eds.) *Plant Physiology* ch. 27. Belmont, California: Wadsworth Publishing Company Inc, 1969.

Kollerstrom, Nick. *Gardening and Planting by the Moon.* Quantum, 2003.

Soper (ed.) *Biological Science* Vol 2 ch. 20. Cambridge University Press, 1984.

Chapter 4

Darwin, C.R. *The Formation of Vegetable Mould Through the Action of Worms with Observations on Their Habits.* John Murray, 1881.

Flowerdew, Bob. *Organic Gardening Booklets: Garden Compost*, Soil Association.

Harris, Dudley: *Illustrated Guide to Hydroponics.* New Holland Publishing, 1994.

Pratt, Mary M. *In Defence of Darwin.* Oikos 31: 349–350, 1978.

Shewell-Cooper, W.E. *Dr Shewell-Cooper's Basic Book of Vegetable Growing*. Granada Publishing (Mayflower Books), 1978.

Chapter 5

Greenwood, Pippa and Halstead, Andrew. *Pests and Diseases*. Dorling Kindersley, 1997. (NB Available pesticides may be out of date)

Roth, Sally. *Weeds, Friend or Foe?* Carroll & Brown, 2002.

Shepherd and Gallant (eds.) *The Little Slug Book*. Centre for Alternative Technology, 2002.

Chapter 6

Baines, Chris. *How to Make a Wildlife Garden*. Elmtree Books, 2001.

Bardsley, Louise. *The Wildlife Pond Handbook.* New Holland for The Wildlife Trusts, 2003.

Chinery, Michael. *Attracting Wildlife to Your Garden.* Collins, 2004.

English Nature. *Gardening with Wildlife in Mind* (CD). The Plant Press. Tel: 01273 476151; e-mail: john@plantpress.com

Hill, Fran. *Wildlife Gardening: A Practical Handbook.* Derbyshire Wildlife Trust, 1988; revised edition, 1996.

Leopold, Aldo. *A Sand County Almanac and Sketches Here and There*. Oxford University Press, 1949 (paperback 1968).

Mabey, Richard. *The Common Ground: A Place for Nature in England's Future.* Hutchinson, 1980.

Owen, Jennifer. *The Ecology of a Garden: The First Fifteen Years*. Cambridge University Press, 1991.

Swift, Katharine. 'Gazing into the Cow Parsley'. *The Times Weekend Review*, 13 June 2004.

Thompson, Ken. 'BUGS in the Borders'. *The Garden*, Journal of the Royal Horticultural Society Vol. 129 pp346–349.

Chapter 7

Coen, E. *The Art of Genes*. OUP, 1999.

Gonick, L. & Wheelis, M. *The Cartoon Guide to Genetics*. Harper Collins, 1991.

Watson, James D. *The Double Helix*. Weidenfeld & Nicolson, 1968.

Chapter 8

Gamlin, Linda and Vines, Gail (eds.) *The Evolution of Life*. Collins, 1987.

Stearn, William T. *Stearn's Dictionary of Plant Names for Gardeners*. Cassell, 1992.

The following titles are recommended for general further reading.

Brickell, Christopher (ed.) *Royal Horticultural Society Encyclopaedia of Gardening.* Dorling Kindersley, 1992.
Buczacki, Stefan. *Understanding Your Garden: The Science and Practice of Successful Gardening.* Cambridge University Press, 1990.
Campbell-Culver, Maggie. *The Origin of Plants.* Headline Book Publishing, 2001.
Ingram, David S., Vince-Prue, Daphne, and Gregory, Peter J. *Science and the Garden.* Blackwell Science for the Royal Horticultural Society, 2002.
Jensen, William A. and Kavaljian, Leroy G. (series eds.) *Plant Physiology.* Belmont, California: Wadsworth Publishing Company Inc, 1969.
Royal Horticultural Society. *The Plant Finder.* Dorling Kindersley, published yearly.
Soper (ed.) *Biological Science.* Vols 1 and 2, Cambridge University Press, 1984.
Spedding, Colin and Geoffrey. *The Natural History of a Garden.* Timber Press, 2003.
Thomas, Joanna. *Gardening Terms Explained.* Weston Publishing [no date]
Thompson, Ken. *An Ear to the Ground: Garden Science for Ordinary Mortals.* Transworld Publishers for Eden Project Books, 2003.

Useful addresses

United Kingdom

www.rhs.org.uk
Royal Horticultural Society. Lots of information for non-members; advice is also available for members from RHS scientific staff. Membership contact details are provided on the site.

www.wildlifetrusts.org
Addresses and information about your local Wildlife Trust.

www.hedgelaying.org.uk
National Hedgelaying Society

www.btcv.org
British Trust for Conservation Volunteers

enquire@hdra.org.uk
Henry Doubleday Research Association, Ryton Organic Gardens, Coventry, Warwickshire CV8 3LG

information@psd.defra.gsi.gov.uk
Information Services Branch, Pesticides Safety Directorate, Mallard House, Kings Pool, 3 Peasholme Green, York YO1 7PX

Original Organics Ltd., Unit 9 Langlands Business Park, Uffculme, Cullompton, Devon EX15 3DA (tel: 01884 841515)
Wormery supplier.

USA

www.ahs.org
American Horticultural Society

www.nwf.org
National Wildlife Federation

www.natureserve.org/explorer/
Equivalent to the UK's National Vegetation Classification, and based on The Nature Conservancy's International Classification of Ecological Communities: Terrestrial Vegetation of the United States

Metric conversion chart

inches	centimetres
⅛	0.3
¼	0.6
⅓	0.8
½	1.25
⅔	1.7
¾	1.9
1	2.5
1¼	3.1
1⅓	3.3
1½	3.75
1¾	4.4
2	5.0
3	7.5
4	10
5	12.5
6	15
7	17.5
8	20
9	22.5
10	25
12	30
15	37.5
18	45
20	50
24	60
30	75
36	90

feet	metres
¼	0.08
⅓	0.1
½	0.15
1	0.3
1½	0.5
2	0.6
2½	0.8
3	0.9
4	1.2
5	1.5
6	1.8
7	2.1
8	2.4
9	2.7
10	3.0
15	4.5
20	6.0
25	7.5
30	9.0
35	10.5
40	12
45	13.5
50	15

Index

Published in 2005 by
Timber Press, Inc.
The Haseltine Building
133 S.W. Second Avenue, Suite 450
Portland, Oregon 97204-3527, U.S.A.

www.timberpress.com

Edited by Sue Viccars
Designed by Les Domintey
Printed in China

ISBN 0-88192-718-X

Catalogue records for this book are available from the Library of Congress
and the British Library.